WHO FOOD ADDITIVES SERIES: 38

Toxicological evaluation of certain veterinary drug residues in food

Prepared by
The forty-seventh meeting of the Joint FAO/WHO
Expert Committee on Food Additives (JECFA)

World Health Organization, Geneva, 1996

IPCS - International Programme on Chemical Safety

The **International Programme on Chemical Safety (IPCS)**, established in 1980, is a joint venture of the United Nations Environment Programme (UNEP), the International Labour Organisation (ILO), and the World Health Organization (WHO). The overall objectives of the IPCS are to establish the scientific basis for assessment of the risk to human health and the environment from exposure to chemicals, through international peer-review processes, as a prerequisite for the promotion of chemical safety, and to provide technical assistance in strengthening national capacities for the sound management of chemicals.

The **Inter-Organization Programme for the Sound Management of Chemicals (IOMC)** was established in 1995 by UNEP, ILO, the Food and Agriculture Organization of the United Nations, WHO, the United Nations Industrial Development Organization and the Organisation for Economic Co-operation and Development (Participating Organizations), following recommendations made by the 1992 UN Conference on Environment and Development to strengthen cooperation and increase coordination in the field of chemical safety. The purpose of the IOMC is to promote coordination of the policies and activities pursued by the Participating Organizations, jointly or separately, to achieve the sound management of chemicals in relation to human health and the environment.

ISBN 92 4 166038 4

CONTENTS

PREFACE

The monographs contained in this volume were prepared at the forty-seventh meeting of the Joint FAO/WHO Expert Committee on Food Additives (JECFA), which met in Rome, Italy, 4-13 June 1996. These monographs summarize the safety data on selected veterinary drug residues reviewed by the Committee. The data reviewed in these monographs form the basis for acceptable daily intakes (ADIs) established by the Committee.

The forty-seventh report of JECFA will be published by the World Health Organization in the WHO Technical Report Series. Terms abbreviated in the monographs are listed in Annex 2. The participants in the meeting are listed in Annex 3 of the present publication and a summary of the conclusions of the Committee is included as Annex 4. Some of the substances listed in this annex were evaluated at the present meeting for residues only.

Residue monographs on the veterinary drugs that were evaluated at the forty-seventh meeting of JECFA will be issued separately by FAO as Food and Nutrition Paper No. 41/9. These toxicological monographs should be read in conjunction with the residue monographs and the report.

Reports and other documents resulting from previous meetings of JECFA are listed in Annex 1.

JECFA serves as a scientific advisory body to FAO, WHO, their Member States, and the Codex Alimentarius Commission, primarily through the Codex Committee on Food Additives and Contaminants and the Codex Committee on Residues of Veterinary Drugs in Foods, regarding the safety of food additives, residues of veterinary drugs, naturally occurring toxicants, and contaminants in food. Committees accomplish this task by preparing reports of their meetings and publishing specifications or residue monographs and toxicological monographs, such as those contained in this volume, on substances that they have considered.

The toxicological monographs contained in this volume are based upon working papers that were prepared by Temporary Advisers. A special acknowledgement is given at the beginning of each monograph to those who prepared these working papers.

Many proprietary unpublished reports are referenced. These were voluntarily submitted to the Committee by various producers of the veterinary drugs under review and in many cases these reports represent the only safety data available on these substances. The Temporary Advisers based the working papers they developed on all the data that were submitted, and all these studies were available to the Committee when it made its evaluations. Special acknowledgement is made to these advisers. The monographs were edited by Dr P.G. Jenkins, International Programme on Chemical Safety.

From 1972 to 1975 the toxicological monographs prepared by JECFA were published in the WHO Food Additives Series; after 1975 this series was available in the form of unpublished WHO documents provided upon request to the Organization.

WHO Food Additives Series No. 20, which was prepared by the twenty-ninth Committee in 1985, through WHO Food Additives Series No. 24, which was prepared by the thirty-third Committee in 1988, were published by the Cambridge University Press. Beginning with WHO Food Additives Series No. 25, which was prepared by the thirty-fourth Committee, WHO has been producing these volumes as priced documents.

The preparation and editing of the monographs included in this volume have been made possible through the technical and financial contributions of the Participating Institutions of the IPCS, which supports the activities of JECFA. IPCS is a joint venture of the United Nations Environment Programme, the International Labour Organisation, and the World Health Organization, which is the executing agency. One of the main objectives of the IPCS is to carry out and disseminate evaluations of the effects of chemicals on human health and the quality of the environment.

The designations employed and the presentation of the material in this publication do not imply the expression of any opinion whatsoever on the part of the organizations participating in the IPCS concerning the legal status of any country, territory, city, or area or its authorities, or concerning the delimitation of its frontiers or boundaries. The mention of specific companies or of certain manufacturers' products does not imply that they are endorsed or recommended by those organizations in preference to others of a similar nature that are not mentioned.

Any comments or new information on the biological or toxicological data on the compounds reported in this document should be addressed to: Joint WHO Secretary of the Joint FAO/WHO Expert Committee on Food Additives, International Programme on Chemical Safety, World Health Organization, Avenue Appia, 1211 Geneva 27, Switzerland.

ADRENOCEPTOR AGONISTS

CLENBUTEROL

First draft prepared by
Dr K.N. Woodward
Veterinary Medicines Directorate
Ministry of Agriculture, Fisheries and Food
Addlestone, Surrey, United Kingdom

1. EXPLANATION

Clenbuterol is a β-adrenoceptor agonist that exerts a potent bronchiolytic effect by preferential action on β_2-adrenoceptors in smooth muscle, resulting in the relaxation of bronchial smooth muscle and a decrease in airway resistance. Similarly, through selective binding to β_2-adrenoceptors on uterine smooth muscle cell membranes, relaxation of the uterus (tocolysis) occurs. Clenbuterol hydrochloride is used for the treatment of respiratory disease in horses and cattle (0.8 μg/kg bw, twice daily) and as a tocolytic agent in cattle (a single parenteral injection of 0.8 μg/kg bw). Although unapproved for use as a repartitioning agent, it is used for this purpose in farm animals at doses several orders of magnitude higher than the recommended therapeutic dose.

Clenbuterol is manufactured as a 50 : 50 racemic mixture, most of its pharmacological activity being associated with the laevo form. It had not previously been evaluated by the Committee. The molecular structure of clenbuterol is shown below.

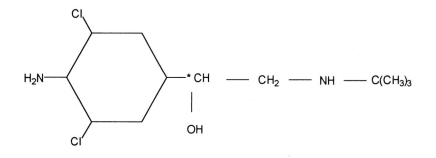

* position of radiolabel in metabolism studies

2.   BIOLOGICAL DATA

2.1   Biochemical aspects

2.1.1   Absorption, distribution and excretion

2.1.1.1   Mice

Following a single intravenous dose of 5 mg/kg bw ^{14}C-clenbuterol to pregnant mice, autoradiography demonstrated high levels of radioactivity in the placenta with lower levels in the fetuses within 30 minutes after administration. Radioactivity was also found in the fetal liver, thymus and spinal column. Total radioactivity in the fetuses was significantly less than in maternal animals (Kopitar, 1969).

2.1.1.2   Rats

After oral administration of 2 mg/kg bw ^{14}C-clenbuterol to rats, approximately 60% of the radioactivity was found in the 48-hour urine sample suggesting good absorption from the gastrointestinal system. A further 20% was recovered from faeces over this time. Around 5-10% was recovered in urine and faeces in the 48-72 hour period. Approximately 7% appeared in the bile. Levels in tissues were relatively low, except for the liver, kidneys and lungs (Kopitar, 1970).

In a separate study using autoradiography, maximum distribution after oral doses of 5 or 10 mg/kg bw occurred approximately 3 hours after administration. Highest concentrations of radioactivity were found in the lungs, liver, kidneys, pancreas and bone marrow, but some radioactivity was also found in the brain, adrenal glands, and skeletal and cardiac muscles (Kopitar, 1969).

When previously untreated rats were given a single oral dose of 5 mg/kg bw clenbuterol by gavage, plasma levels of the drug were found to be 3-5 times lower than in rats pre-treated with the drug at 5 mg/kg bw per day for 6 months. The single dose was found to abolish gastrointestinal peristalsis in the rat but after 6 months of compound administration, peristalsis returned. This may explain the lower, but prolonged plasma levels after the single administration, which probably led to a depot effect in the gut (Kopitar & Zimmer, 1973).

Clenbuterol readily enters the placenta in the pregnant rat. At 3 hours after an intravenous or oral dose of ^{14}C clenbuterol, there was more radioactivity in the placenta than in the blood and tissues of maternal animals or fetuses (Richter, 1982).

2.1.1.3   Dogs

When beagle dogs were given capsules containing ^{14}C-clenbuterol at a dose of 2.5 mg/kg bw, maximum blood and plasma levels occurred 8 hours after dosing. Around 85% of the dose was recovered in the urine after 96 hours, with 4-9% in the corresponding faecal samples. Pretreating dogs for 5 weeks with 2.5 mg/kg bw per day clenbuterol led to a slight increase in the rate of absorption in a manner similar to that noted in rats (Zimmer, 1974a).

After single oral doses of 2.5 mg/kg bw, blood concentrations in the dog were approximately twice those of rabbits given identical doses (Zimmer, 1974b).

Clenbuterol was found to cross the placenta and enter the fetuses in the dog when a single pregnant female was given an oral dose of 2.5 mg/kg bw ^{14}C-clenbuterol. At 4 hours after administration the concentration of radioactivity in fetal plasma was around 16% of that in maternal plasma. Approximately 0.4% of the total dose administered to the maternal animal was recovered in the fetuses at 4 hours after administration (Rominger & Schrank, 1982).

2.1.1.4   Monkeys

After intravenous administration of 100 μg ^{14}C-clenbuterol per animal (approximately 30 μg/kg bw) to cynomolgus monkeys, around 60%

of the dose was recovered in the urine and 4% in the faeces within 72 hours of administration. There was biphasic elimination. It was not possible to calculate the half-life for the first and rapid phase, but the second phase half-life was 20-30 hours (Chasseaud *et al.*, 1978).

2.1.1.5   Baboons

Following a single intravenous dose of 2.5 mg/kg bw ^{14}C-clenbuterol to male baboons, around 82% of the administered dose was recovered in the urine and faeces within 5 days, with the majority of this (72%) within 48 hours. Of the 82%, approximately 68% was recovered in the urine and 14% in the faeces. A similar pattern was noted in baboons given the drug orally. There was a rapid distribution phase and the highest levels of radioactivity were found in the lungs, liver and kidneys (Schmid, 1982).

Placental transfer of ^{14}C-clenbuterol was investigated in a single pregnant baboon given 3.3 mg/kg bw intravenously. The animal was killed 3.5 hours after dosing, and around 1.5% of the total administered dose was found in the fetus (Schmid, 1980).

2.1.1.6   Cattle

In cattle given 0.8 μg/kg bw ^{14}C-clenbuterol using a single intra-muscular injection, the liver was found to have the highest concentrations of radioactivity. In muscle, soon after administration, the radioactivity was due almost entirely to parent compound (Schmid, 1990a).

In a pilot study with a single cow given 0.8 μg/kg bw ^{14}C-clenbuterol orally, the first phase plasma elimination half-life was 0.1 hours and the second phase elimination half-life 17 hours. The maximum plasma concentration was 0.4 μg/ml with t_{max} at 8.5 hours. Approximately 81% of the dose was excreted in the urine with 13% in the faeces over the 96-hour period after dosing; around 1% was excreted in the milk (Schmid & Zimmer, 1977a). Similar results were noted after intramuscular (Schmid & Zimmer, 1977b) and intravenous (Schmid, 1977) administration, but with the latter route plasma levels rose much more rapidly. In both cases, radioactivity was found in the faeces suggesting biliary excretion.

When calves were given 0.8 μg/kg bw ^{14}C-clenbuterol intra-muscularly, twice a day for 2 days and then orally twice a day for 3 days in a consecutive manner, using a combination product also containing trimethoprim and sulfadiazine, around 64% of the dose was excreted during 15 days in urine and 5% in the faeces. The highest level of radioactivity was found in injection site muscle, liver and kidney (Hawkins *et al.*, 1985a). Similar results were seen in another study where cattle were given intramuscular injections of ^{14}C-clenbuterol (Cameron & Phillips, 1987a).

In cattle given 21 intramuscular doses of ^{14}C-clenbuterol at a dose level of 0.8 μg/kg bw per day, twice daily, the majority of the dose was recovered in the urine. The plasma half-life following the last dose was of the order of 4 days. Highest levels of radioactivity were found in liver and kidney, with much lower levels in muscle and fat (Hawkins *et al.*, 1993a).

After administration of 7 μg/kg bw ^{14}C-clenbuterol by the intramuscular route to a pregnant cow, radioactivity crossed the placenta and was found in fetal tissues, including skeletal tissues, kidneys and liver, when the animal was killed 3 hours after the dosing. Radioactivity was also found in the maternal and fetal eyes (Baillie *et al.*, 1980).

When lactating cattle were given oral or intramuscular doses of ^{14}C-clenbuterol at a dose of 0.8 μg/kg bw, radioactivity was found in the urine for 48-70 hours after administration. In the first 48 hours, the majority of the radioactivity in milk was attributable to parent compound (Schmid, 1990b).

Lactating dairy cows treated with 0.8 μg/kg bw ^{14}C-clenbuterol, twice daily by the intramuscular route for 3 consecutive days, and twice daily by the oral route for a further 2 days followed by a single oral dose on day 6, were found to have milk concentrations in the range of 0.76-1.03 μg/litre 12 hours after the first dose rising to 2.14-2.6 μg/litre after 48 hours. They remained in the range 1.46 to 3.93 μg/litre throughout the dosing period. Highest tissue concentrations were at the injection site, liver and kidney (Hawkins *et al.*, 1985b). Similar results were obtained in another study where dairy cattle were given oral, intravenous or intramuscular doses of 0.8 μg/kg ^{14}C-clenbuterol. In this study approximately 2% of administered radioactivity was found in milk (Cameron & Phillips, 1987b).

2.1.1.7   Horses

After administration of a single oral dose of 0.8 μg/kg bw ^{14}C-clenbuterol to a horse, the maximum plasma concentration of 0.4 μg/ml was reached at approximately 2 hours. The elimination half-life was 21 hours. After 144-168 hours, 81% of the administered dose was recovered in urine. Following intravenous administration, 82% of the administered dose was found in urine, with around 9% in faeces, suggesting biliary excretion (Zimmer, 1977).

In a study with 10 horses, animals were given a single oral dose of 0.8 μg/kg bw clenbuterol followed by a single intravenous dose 3 weeks later. A second group of 10 horses was treated in a similar manner, but initially using the intravenous route with the oral dose after 3 weeks. Approximately 20-50% of the oral dose (mean 36%) as parent drug, was excreted over 108 hours in the urine, while, following intravenous administration, some 17-70% of the dose (mean 43%) was excreted in the 108-hour urine (Zupan, 1985).

In a study with a clenbuterol, trimethoprim and sulfadiazine combination product, horses were given gelatin capsules containing ^{14}C-clenbuterol at a dose of 0.8 µg/kg bw, twice daily for 10 days, with a single oral dose on day 11. The majority of the dose (70-77%) was excreted in the urine and 10-15% in the faeces. The maximum plasma concentration was achieved 3-4 hours after the first administration and the plasma half-life was 9-10 hours. The mean plasma concentration was 0.6-1.2 µg/ml. Highest levels of radioactivity were found in the liver and kidneys with very little in fat (Hawkins *et al.*, 1984).

Similarly, when 12 horses were given 21 twice daily oral doses of 0.8 µg/kg bw ^{14}C-clenbuterol, steady-state plasma levels were rapidly achieved. Peak levels of 1-1.9 ng equivalents/ml occurred 1-3 hours after dosing. The terminal half-life was 22 hours. Around 75-91% of the dose was excreted in the urine and 6-15% in the faeces. When the horses were killed 3 or 12 hours and 9, 12 or 28 days after dosing, the highest levels of radio-activity were found in the liver and kidney. Levels in fat were extremely low (Johnston & Dunsire, 1993).

2.1.2   Biotransformation

The biotransformation of clenbuterol in rats, rabbits and dogs is complex. Five urinary metabolites have been identified in dogs, and eight in the rat and rabbit. Unchanged clenbuterol is the major compound found in the urine of these species, and a range of oxidative and conjugated metabolites are formed. Of these, 4-amino-3,5-dichloromandelic acid, 3-amino-3,5-dichlorobenzoic acid and 4-amino-3,5-dichlorohippuric acid are the major components. In the baboon, unchanged parent drug is the major urinary component (18%), again with a range of metabolites (Zimmer, 1971, 1974b; Johnston & Jenner, 1976; Schmid, 1982; Schmid & Prox, 1986).

In cattle, the major compound found in urine (28-52%) and in the liver (50-80%) after administration of clenbuterol was the parent compound. Small amounts of 4-amino-3,5-dichlorobenzoic acid were found in urine and liver, and 4-amino-3,5-dichlorohippuric acid was detected in urine (Baillie *et al.*, 1980; Hawkins *et al.*, 1993a). In muscle, the main component was parent compound (Schmid, 1990a).

In horses also, parent clenbuterol was the major compound found in urine (30-50%). It was also the major component in liver and kidney (Zimmer, 1977; Hawkins *et al.*, 1984; Johnston & Dunsire, 1993). A study to examine specifically the nature of the metabolites in horse liver confirmed clenbuterol as the major component, but with a smaller quantity of a metabolite identified as 1-(4-amino-3,5-dichlorophenyl)-1,2-ethanediol (Hawkins *et al.*, 1993b).

None of the studies with any species revealed evidence of significant covalent binding of clenbuterol or its metabolites.

Overall, the data suggest that the metabolism of clenbuterol is similar in all the species studied and that the major differences are quantitative rather than qualitative. The proposed metabolic pathways are shown in figure 1.

2.1.3   Kinetics in humans

Following a single oral dose of 20 μg ^{14}C-clenbuterol hydrochloride to volunteers, 67% of the administered dose was excreted in the urine; faecal excretion was very low. The major compound present in the urine was unchanged parent drug. The metabolic profile was similar to that noted in animals (Zimmer, 1974c).

When a single dose of 20 μg ^{14}C-clenbuterol hydrochloride was administered orally to volunteers, the highest plasma level was 0.11 μg/litre, with a t_{max} of 2-3 hours. Around 87% of the administered radioactivity was detected in the urine; correspondingly lower levels of radioactivity were found in the faeces.

In a repeat-dose phase of the same study, volunteers were given 40 μg ^{14}C-clenbuterol for 2 days and then 20 μg for a further 2 days. Around 75% of the dose was detected in the urine. The metabolite profile was similar to that noted in animals, and the major urinary component was unchanged parent drug; the terminal half-life was around 34 hours (Zimmer, 1976).

Oral doses of 80 μg clenbuterol hydrochloride were given to 12 pregnant women in premature labour. These were followed 12 hours later by 2 oral doses of 40 μg, and then by maintenance doses of 20 μg, generally at 12-hour intervals. Blood samples were taken on days 2, 4, 6 and 8 of study, 3 hours after the morning dose. A steady-state average concentration of 0.298 μg/litre was achieved, with a minimum of 0.28 and a maximum of 0.344 μg/litre. Bioavailability was > 80% (Rominger *et al.*, 1987).

There are no data on *in vivo* percutaneous absorption of clenbuterol in humans. However, data using a model which utilizes excised human skin (Rohr *et al.*, undated) suggests that some absorption can occur (Boehringer, 1991b).

Figure 1. Biotransformation of clenbuterol; general aspects in a number of
mammalian species

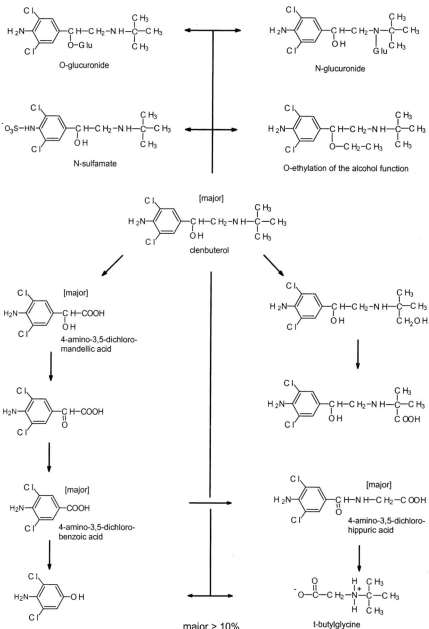

major > 10%

2.2 Toxicological studies

2.2.1 Acute toxicity studies

The results of acute toxicity studies after administration of clenbuterol are summarized in Table 1.

Table 1. Acute toxicity of clenbuterol

Species	Sex	Route	LD_{50} (mg/kg bw)	Reference
Mouse	M	oral	80	Nishikawa et al., 1972
	F	oral	133	Nishikawa et al., 1972
	M & F	s.c	72	Nishikawa et al., 1972
	M & F	i.v	42	Nishikawa et al., 1972
	M	i.p	46	Nishikawa et al., 1972
	M	i.p	72	Nishikawa et al., 1972
Rat	M & F	oral	175	Nishikawa et al., 1972
	M	oral	82	Nishikawa et al., 1979
	F	oral	170	Nishikawa et al., 1979
	M & F	i.v	30	Nishikawa et al., 1972
	M	i.v	35	Nishikawa et al., 1979
	F	i.v	39	Nishikawa et al., 1979
	M & F	s.c	170	Nishikawa et al., 1972
	M & F	i.p	70	Nishikawa et al., 1972
	M & F	oral[1]	> 0.125	James, 1990a
Rabbit	M & F	i.v	84	Nishikawa et al., 1972
	M & F	dermal[1]	> 0.05	James, 1990b
Dog	M & F	oral	400-800	Pappritz, 1968
	M & F	i.v	45-52	Pappritz, 1972

[1] Administered as a proprietary veterinary medicinal product containing 25 μg clenbuterol per ml

Signs of toxicity in mice, rats and rabbits included lethargy, increased heart rates, and tonic-clonic convulsions. After intravenous dosing, death, when it occurred, did so either during injection or immediately after it. In dogs, all the oral doses employed (400-800 mg/kg bw) caused increases in heart rate. Salivation, lethargy and tonic-clonic convulsions also occurred.

In cattle, single intravenous doses of up to 3.6 μg/kg bw resulted in increases in heart and respiratory rates (Johnson, 1987). Single oral doses of 2.4-12.0 μg/kg bw in horses resulted in sweating and elevated heart rates and body temperatures at all dose levels (Hamm & Erichsen, 1989).

2.2.2 Short-term toxicity studies

2.2.2.1 Mice

Groups of 15 male and 15 female ICR-JCL mice were given daily oral doses of 0, 2.5, 12.5 or 62.5 mg/kg bw clenbuterol hydrochloride as an aqueous solution, by stomach tube, for 30 days. Body weight gain and food consumption were increased in all treated groups in a dose-related manner. There were no adverse effects on haematology. The animals from the highest dose group appeared weak and 6 animals died during the course of the study.

At necropsy, there were significant increases in liver weights in animals given 12.5 or 62.5 mg/kg bw per day. Ischemic heart damage was noted in one animal given 12.5 mg/kg bw per day, and in two animals from the highest dose group. The NOEL in this study was 2.5 mg/kg bw per day (Kast, 1973a).

2.2.2.2 Rats

Groups of 15 male and 15 female Sprague-Dawley JCL rats were given daily oral doses of 0, 1, 10 or 100 mg/kg bw per day clenbuterol hydrochloride in distilled water, by stomach tube, for 30 days. Signs of toxicity were seen at the highest dose and these included salivation, encrusted eyes and noses, and reduced body weight gain and food consumption; 50% of these animal died during the course of the study. Serum aspartate aminotransferase and glucose concentrations were significantly reduced in animals of both sexes given 10 or 100 mg/kg bw per day.

Liver weights were significantly reduced in males given the highest dose, while ischaemic lesions were seen in cardiac muscle in one male and one female given this dose. The NOEL in this study was 1 mg/kg bw per day based on blood chemistry findings (Kast & Tsunenari, 1973a).

Groups of 15 male and 15 female Sprague-Dawley JCL rats were given doses of 0, 1, 5, 25 or 75 mg/kg bw per day clenbuterol hydrochloride in the feed for 6 months. At the highest dose level, body weight gain and food consumption were reduced. During the first week, a reduction in water consumption occurred in the animals given this dose, and nine male and five female animals died during the course of the study.

At necropsy, ischaemic lesions of the myocardium were noted in all treated animals and there was a dose-related incidence and severity. These lesions were also found in another group given 75 mg/kg bw per day clenbuterol hydrochloride and then allowed a 4-week recovery period without drug treatment. A NOEL was not identified in this study (Kast & Tsunenari, 1973b).

Groups of 20 male and 20 female FW49/Biberach rats were fed diets containing clenbuterol hydrochloride for 18 months. Further groups of three male and three female rats were fed treated diets for 14 days, followed by a recovery period; these animals were the subject of ECG investigations. Another group of 10 female and 10 male rats was fed the highest dietary level and then maintained for a 6-week recovery period. The overall design of the study, including doses, is set out below.

Dose (mg/kg bw per day)	18-month phase number of animals		14-day ECG study[2] numbers of animals	
	males	females	males	females
0	20	20	3	3
0.1	24	20	3	3
0.5	20	20	3	3
2.5	20	20	3	3
2.5	10[1]	10[1]	-	-

[1] subject to 6 week recovery period
[2] subject to 21 day recovery period

Body weight gain in treated groups was slightly increased when compared with control values. However, there were no overt signs of toxicity, and no deaths occurred. There were no significant haematological effects. However, there were reductions in myocardial glycogen, skeletal muscle glycogen and liver glycogen values in all treated groups, but there was no clear dose response. Histochemical examination of the myocardium showed a reduction in enzyme activity, particularly in the lowest dose group.

The ECG examination revealed a bradycardia rather than the expected tachycardia, but no explanation was provided for this. Heart rates returned to normal 2 days into the recovery period. As the results of this study appear to conflict with those of other studies (reduction in heart rate rather than the expected increase), and as the hearts were not adequately examined macroscopically or microscopically, an NOEL cannot be identified from this study (Serbedija & Bauer, 1973).

Groups of 15 male and 15 female Sprague-Dawley JCL rats were given daily intravenous doses of 0, 1, 4 or 16 mg/kg bw clenbuterol hydrochloride for 30 days. Overt signs of toxicity included weakness at the two highest dose levels and increased respiratory and heart rates at the highest dose level. One female and one male in the 1 and 4 mg/kg bw per day dose groups respectively, died, while five male and five female animals died at the highest dose. There was evidence of hepatotoxicity (focal necrosis) in rats given the highest dose. A NOEL was not identified in this study (Kast & Tsunenari, 1973c).

Groups of 30 male and 30 female Chbb: THOM (SPF) rats were exposed by the inhalation route to 0 or 0.667 mg/m^3 clenbuterol hydrochloride in aqueous solution. A further two groups of 15 male and 15 female rats were exposed to 5.013 or 40.0 mg/m^3. The exposures were conducted for 0.5-2 hours per day, 7 days per week, for 13 weeks. These exposures gave corresponding doses of 0, 0.01, 0.16 and 2.58 mg/kg bw per day. Groups of 10 male and 10 female control animals, and 10 male and 10 female rats exposed to the highest concentration were subjected to a 6-week recovery period without further exposure to clenbuterol hydrochloride.

There was no drug-related mortality during the study. Increased heart rates were noted in all treated rats. Animals exposed to the highest concentration of clenbuterol hydrochloride showed myocardial necrosis. Myocardial changes were noted in animals subject to the 6-week recovery period. As increases in heart rate were noted at all concentrations, a NOEL was not identified in this study (Koellmer *et al.*, 1977).

Groups of 20 male and 20 female Charles River CD rats were exposed to aerosols of clenbuterol hydrochloride in ethanol with dichloro-fluoromethane and 1,2-dichlorotetrafluoroethane as propellants, using head-only exposure. Other groups served as controls or ethanol/propellant controls. The doses were equivalent to up to 118 μg/kg bw per day.

The only treatment-related effects attributable to clenbuterol hydrochloride were liver weight reduction and small aggregations of alveolar macrophages, which were clearly dose-related (Clark *et al.*, 1979).

However, there were insufficient data on dosimetry to identify a NOEL and in addition, there were no measurements of heart-rate or other cardiac function parameters.

2.2.2.3   Dogs

Groups of three male and three female beagle dogs were given starch capsules containing clenbuterol hydrochloride at doses of 0, 0.4, 4 and 20 mg/kg bw per day, daily for 13 weeks. The highest dose was increased to 30 mg/kg bw per day from week 5 and to 40 mg/kg bw per day from week 9. Water consumption was normal during the study, but a slight decrease in appetite was noted in high-dose animals when the dose was increased to 40 mg/kg bw per day. The rate of body weight gain significantly fell from this point. One male and two females given the highest dose appeared sedated after treatment. One of these became increasingly agitated and breathless and died 2-3 minutes after receiving the 23rd dose of 20 mg/kg bw per day. Tachycardia occurred after the first dose in all treated animals and this persisted until the following day. There was a decrease in its severity during subsequent weeks, but it continued to be noted in all treated animals. A shortening of the P-Q, Q-T and T-P intervals in the ECG, corresponding to the increased heart rates, was seen in all treated dogs. The dog that died was found to have a severe acute pulmonary infection. Myocardial haemorrhage and necrosis, damage to the hepatic parenchyma and severe haemorrhagic bronchopneumonia were noted. No abnormalities occurred in any of the other animals. As dose-related tachycardia occurred at all doses, a NOEL was not identified in this study (Leuschner et al., 1969).

Groups of three male and three female beagle dogs were given oral doses of 0, 2.5 and 40 mg/kg bw per day clenbuterol hydrochloride in gelatin capsules for 13 weeks. No animals died during the study, but in the first week, all dogs given clenbuterol hydrochloride were apathetic, had redness of the skin and eyes, mydriasis and fixed pupils. Tachycardia was seen in all dogs. At 2.5 and 40 mg/kg bw per day, significant increases in the activities of several serum enzymes, including LDH and creatinine kinase, were found. The glycogen levels of the left heart muscle were decreased. Myocardial necrosis was seen in dogs given the 2.5 and 40 mg/kg bw per day doses. A NOEL was not identified in this study (Pappritz & Bauer, 1973a).

Groups of three male and three female beagle dogs were given gelatin capsules containing daily doses of 0, 0.1 or 0.5 mg/kg bw per day clenbuterol hydrochloride for 1 year. A group of six male and six female dogs were given 2.5 mg/kg bw per day in the same manner, and three animals of each sex were maintained on a control diet for 6 weeks at the end of the study, as part of a recovery phase. No animals died during the study. In the first week all drug-treated animals were apathetic and had diffuse reddening of the skin. Tachycardia was noted in all treated animals, as was mydriasis, episcleral vascular injection and conjunctivitis.

There were no significant effects on food consumption or body weights, although dogs given the highest dose gained slightly more weight than did those in other groups. No haematological effects were seen, but there were increases in some serum enzymes in treated groups. Liver glycogen was decreased in dogs given the highest dose and this remained decreased at the end of the 6-week recovery period.

Heart weights were increased in all the treated animals when compared with controls, and myocardial necrosis was noted in the papillary muscle of the left ventricle of all these dogs, including those subject to the

6-week recovery period. There was no dose relationship in the cardiotoxicity seen, but histochemical methods revealed smaller decreases in the activities of several enzymes in cardiac muscle of the 0.1 mg/kg bw per day group than in the other treatment groups. A NOEL was not identified in this study (Pappritz & Bauer, 1973b).

Groups of three male and three female beagle dogs were given daily intravenous injections of 0, 1, 10 or 1000 μg/kg bw per day clenbuterol hydrochloride for 30 days. Dogs given the highest dose of clenbuterol hydrochloride were lethargic and all treated animals developed tachycardia with associated decreases in the P-Q and Q-T intervals in the ECG. At necropsy, one male given the highest dose was found to have myocardial necrosis. A NOEL was not identified in this study (Pappritz *et al.*, 1973).

2.2.2.4   Monkeys

Groups of three male and three female cynomolgus monkeys were exposed to atmospheric levels of clenbuterol hydrochloride as an aerosol in ethanol, with dichlorofluoromethane and 1,2-dichlorotetrafluoroethane as propellants, using an oropharyngeal tube, giving doses of 25, 50 and 150 μg/kg bw per day for 26 weeks. Half of the daily metered dose was given each morning and the remainder in the afternoon. Another group of three male and three female monkeys served as sham controls with no exposure, while a further group of three male and three female monkeys served as placebo controls and were exposed to the same aerosol as treated animals, but without the clenbuterol hydrochloride.

No signs of toxicity were seen in this study and food consumption and body weights were normal except in high-dose females where a large body weight gain occurred. Ophthalmoscopy, haematology, lung mechanics and ventilation, blood chemistry and urinalyses were normal. A low oxygen tension was noted in high-dose animals in week 10 of the study. There were no drug-related changes in heart rate or ECGs. At macroscopic and histo-pathological examination, there were no drug-related changes (Collins *et al.*, 1979).

Thus, the NOEL in this study was of the order of 25 μg/kg bw per day by the inhalation route, but this could not be identified as a definitive study because of the problems in dosimetry in an inhalation study of this type. For example, there were no data on how much drug reached the lower respiratory tract and no information regarding blood levels. Hence, the degree of systemic exposure was unknown.

2.2.2.5   Cattle

In calves given the therapeutic dose of 0.8 μg/kg bw per day clenbuterol hydrochloride for 10 days by the oral or intravenous routes, only transient tachycardia occurred. When given at 5 times the therapeutic dose

(4 μg/kg bw per day) for 10 days, tachycardia and audibly increased heart force were the only effects noted. When given to adult cattle at the therapeutic dose or twice the therapeutic dose for periods of up to 9 days, by the oral route, the main effects were tachycardia and falls in diastolic blood pressure (Fenner, 1982; Cameron *et al.*, 1992).

2.2.2.6   Horses

No clinically significant adverse effects were noted in horses given 0.8-17.5 μg/kg bw per day clenbuterol hydrochloride by the oral route for periods of up to 64 days. Transitory effects included tachycardia, sweating, muscle tremors, hyperglycaemia and hypophosphataemia (Erichsen, 1988, 1989; Hamm & Erichsen, 1989; Owen *et al.*, 1990).

Clenbuterol, like other β-agonists, leads to tachycardia and hypotension. This probably results in reduced myocardial perfusion at a time when oxygen demand is high because of increased cardiac rate. The end result is hypoxia which probably leads to the necrotic lesions seen in the left ventricular papillary muscles (Rosenblum *et al.*, 1965; De Busk & Harrison, 1969; Poynter & Spurling, 1971; Roberts & Cohen, 1972; Magnusson & Hansson, 1973; Balazs & Bloom, 1982). This explains the cardiac effects noted in the repeated dose (and other) studies with clenbuterol.

2.2.3   Long-term toxicity/carcinogenicity studies

2.2.3.1   Mice

Groups of 50 male and 50 female (C57B/6 x DBA/2) mice were given clenbuterol in the drinking-water at doses of 0.1, 1.0 and 25.0 mg/kg bw per day for 2 years. Groups of 100 males and 100 females were given drinking-water only over this period and served as controls. Treated animals consumed slightly more food and water than controls, but showed no overt signs of toxicity, except for a dose-related increase in body weight in the first year and a decrease during the second year. Survival after 2 years was good in males (78-86%) and acceptable in females (54-60%).

At termination, there were significant changes in the weights of a number of organs when compared with control values. These included significant increases in the absolute heart weights of males given 1 mg/kg bw per day and in females given 25 mg/kg bw per day. There were significant increases in relative heart weights in all treated male and female mice. The incidences of non-neoplastic lesions, including myocardial lesions, were not treatment related. There was no increased incidence of any tumour type in treated mice when compared with control values (Umemoto, 1984).

2.2.3.2   Rats

Groups of 48 male and 48 female Chbb:THOM rats were fed diets containing clenbuterol hydrochloride, resulting in daily doses of 6.25, 12.5, or 25.0 mg/kg bw per day for 2 years. A further group of 72 rats of each sex was fed a diet containing no clenbuterol hydrochloride. As β-sympathomimetic agents are known to induce mesovarian leiomyomas in some strains of rat, groups of 72 male and 72 female Charles River Sprague-Dawley rats were also fed diets containing clenbuterol hydrochloride at doses of 0 or 25 mg/kg bw per day for 2 years. After one control rat began to show signs of clenbuterol-related effects, contamination of the feed preparation personnel and equipment was confirmed, and the substance was then administered via the drinking-water to ensure the same daily doses.

Treated rats appeared nervous, tense and aggressive, but no effects on mortality were seen. Survival was in the range of 70-87% in males and 50-71% in females. In both sexes and strains body weight was reduced in a dose-related manner. Consumption of drinking-water was significantly reduced in SD rats given the highest dose. At termination, fibrosis of the subendocardial connective tissue was frequently observed in all groups including controls, but the incidence was higher in the Chbb:THOM rats, suggesting a drug-related effect.

There was an increased incidence of mesovarian leiomyomas in the female Sprague-Dawley rats, but not in the Chbb:THOM rats. The incidence is shown in Table 2.

Table 2.   Incidence of mesovarian leiomyomas in female rats treated with clenbuterol hydrochloride for 2 years

	Chbb:THOM rats				Sprague-Dawley rats	
Dose (mg/kg bw per day)	0	6.25	12.5	25.0	0	25.0
Numbers with mesovarian leiomyomas per group	0/72	0/48	0/48	0/72	0/72	11/72

There was no increased incidence of any other tumour type (Serbedija *et al.*, 1982).

Leiomyomas of the smooth muscle of the uterus in mice and mesovaria of rats have been reported following long-term treatment with β-agonists (Nelson & Kelly, 1971; Jack *et al.*, 1983; Amemiya *et al.*, 1984; Gibson *et al.*, 1987; Gopinath & Gibson, 1987; Sells & Gibson, 1987). They are not associated with genotoxicity with these agents and are considered to be associated with adrenergic stimulation. The β-agonist salbutamol produced mesovarian leiomyomas in rats, but these could be prevented by administration of the β-blocking agent propranolol; a similar phenomenon has been reported in mice where the induction of leiomyomas of the uterus by medroxalol was prevented by propranolol (Jack *et al.*, 1983; Gibson *et al.*, 1987). There have been no reports of increased incidences of leiomyomas in women following the use of β-adrenergic agents (Poynter *et al.*, 1978).

2.2.4   Reproductive toxicity studies

2.2.4.1   Rats

Groups of 17-20 pregnant female Chbb:THOM rats were given daily oral doses of 0, 1, 7 or 50 mg/kg bw per day clenbuterol hydrochloride in distilled water, by gavage, from day 15 of gestation to 21 days post-partum when dams and pups were sacrificed. The pups, including those that died before day 21, were necropsied and X-rayed. Maternal body weights were reduced at the end of the study in animals given 7 mg/kg bw per day clenbuterol hydrochloride. Those given the highest dose were similar to the controls, but these animals were heavier at the beginning of the study. Maternal food consumption was reduced in all treated groups.

The number of pups born dead increased in a dose-related manner, as did the number of pups dying after birth. At the highest dose all the pups died, as shown below.

Dose level	Control bw per day	1 mg/kg bw per day	7 mg/kg bw per day	50 mg/kg bw per day
Number of pups born alive	204	173	153	103
Number of pups born dead	7 (3%)	11 (6%)	31 (17%)	59 (36%)
Number of pups dying after birth	4 (2%)	21 (12%)	68 (44%)	103 (100%)

Body weights of pups were also reduced in a dose-related manner. A NOEL was not identified in this study (Lehman, 1974).

Groups of 30 male and 30 female Chbb:THOM rats were given daily oral doses of 0, 1, 7 or 50 mg/kg bw per day clenbuterol hydrochloride in distilled water by gavage. Treatment of the male rats commenced 10 weeks prior to mating while that of females began 2 weeks prior to mating. On day 14 of gestation 50% of the females were killed and the uterine contents examined. The remaining female rats were allowed to deliver normally and rear their pups until weaning. Dosing of all females continued until they were killed on day 14 of gestation or after weaning.

To investigate the cause of the high pup mortality in the study of Lehman (1974), the litters of six control dams were exchanged with litters from dams given 50 mg/kg bw per day. Surviving dams and pups were necropsied at the end of the lactation period. Food consumption was increased in treated animals when compared with control values. However, body weight was significantly reduced in animals of both sexes given the highest daily doses.

Pup weights at birth were reduced in all treated groups. There was a dose-related decrease in the numbers of viable pups which was statistically significant at 7 and 50 mg/kg bw per day. All the pups in the 50 mg/kg bw per day group died on the first day of lactation regardless of whether or not they were suckled by treated or control dams. The majority of pups from control dams suckled by dams that were given the 50 mg/kg bw per day dose survived.

There were no substance-related effects on fertility, corpora lutea, implantation or resorption rates, and no malformations were noted in pups from treated rats. The hearts of three pups from high-dose females were examined histologically, but no abnormalities were found (Lehman, 1975).

Groups of 34 male and 34 female Chbb:THOM rats were given daily oral doses of 0, 1.5, 7.5 or 15 μg/kg bw per day clenbuterol hydrochloride in distilled water, by gavage. The male rats were treated for 70 days prior to mating and the females from 14 days prior to mating up until the end of gestation. The pups were not treated. Dosing of the females continued until the interim sacrifice or after weaning.

On days 13-15 of gestation, 50 pregnant rats were killed and the uterine contents examined. The remainder were allowed to give birth naturally and the pups reared until 3 weeks old, except for two males and two females in each litter which were reared until 10 weeks after mating when certain behavioural tests were conducted (pupillary reflex, photophobia response, hearing, behaviour on a rotating rod, behaviour in a Y maze). The animals were then mated and allowed to litter naturally.

There were no effects on fertility attributable to clenbuterol treatment, and gestation length, numbers of corpora lutea, implantation rates, incidences of resorptions, parental body weights and food consumption were unaffected.

No compound-related effects were seen in the littering phase and the pups gained weight normally. Performance in the behavioural tests were similar in treated and control groups and the offspring of these animals were normal in all respects. The NOEL was 15 μg/kg bw per day (Lehman, 1980).

Similar results were obtained in an almost identical study on rats using the same dose levels. Again, the NOEL was 15 μg/kg bw per day (Lehman, 1981).

No teratogenic effects were seen in any of these reproductive toxicity studies.

2.2.5   Special studies on embryotoxicity/teratogenicity

2.2.5.1   Rats

Groups of 20 mated female SPF-FW 49 Biberach rats were given oral doses of 0, 0.04, 0.2 or 1 mg/kg bw per day clenbuterol hydrochloride in distilled water, by gavage on days 6-15 of gestation. No evidence of maternal toxicity was seen and there were no compound-related effects on the incidences of resorptions, or on litter and fetal weights. There were no increased incidences of any type of malformation. The NOEL was 1 mg/kg bw per day (Lehman, 1969a).

Groups of 25 mated female Sprague-Dawley rats were given oral doses of 0, 0.01, 1, 10 or 100 mg/kg bw per day clenbuterol hydrochloride in distilled water, by gavage, from days 9 to 14 of gestation. Approximately 24 hours before the expected delivery date, 20 dams per dose level were killed and the uterine contents examined while the remaining five pregnant animals in each group were allowed to deliver naturally.

Overt signs of toxicity, including weakness, hypersensitivity and bloody vaginal discharge were seen in rats given the highest daily dose. Maternal body weights were reduced in dams given the 10 or 100 mg/kg bw per day doses. Animals given these doses had significantly increased incidences of resorptions with corresponding reductions in the numbers of viable fetuses. Mean litter weights were reduced at both these doses, while mean fetal weight was significantly reduced at the highest dose level.

The incidences of malformations were significantly increased at the two highest dose levels used. Anomalies included hydrocephalus, anasarca, umbilical hernia, anophthalmia, rib variations and splintering of the vertebrae (Kast, 1973b).

A study was conducted to verify the results of Kast (1973b). Groups of 25 mated Sprague-Dawley rats were given 0, 0.01, 1, 10 or 100 mg/kg bw per day clenbuterol hydrochloride in distilled water, by gavage, from days

8 to 17 of gestation. Approximately 24 hours before the expected delivery date, 20 dams per group were killed and the uterine contents examined. The remainder were allowed to deliver naturally.

Overt signs of toxicity were similar to those in the previous study and five dams given the highest dose died. The incidences of resorptions were increased in animals given the highest dose; mean litter weights and fetal weights were reduced. The incidences of malformations were significantly increased at the two highest doses. The types of malformation were similar to those seen in the study of Kast (1973b).

In pups derived from animals allowed to deliver naturally, body weights were increased at the highest dose. Swimming function was retarded in this group. At 10 mg/kg bw per day, two pups were found to have hydrocephalus while at the highest dose level, four pups were found to have anophthalmia. No malformations were noted in pups from the 0.01, 0.1 or 1 mg/kg bw per day groups. The NOEL in both these studies was 1 mg/kg bw per day (Kast, 1975).

Groups of 20 Sprague-Dawley rats were exposed nose only to an atmosphere containing clenbuterol hydrochloride during days 6-15 of gestation. The estimated doses were 0, 19, 39 and 78 µg/kg bw per day. There was no evidence of maternal toxicity or teratogenicity, but there were dose-related increases in the incidences of skeletal variations, indicative of fetotoxicity, at all dose levels (Palmer *et al.*, 1978).

2.2.5.2   Rabbits

Groups of 15 mated Russian/Biberach rabbits were given daily oral doses of 0, 30, 100 or 300 µg/kg bw per day clenbuterol in distilled water, by gavage, from days 6 to 18 of gestation. Three dams given the 300 µg/kg bw per day dose and one given the highest dose died during the study. These deaths were due to bronchopneumonia. Several rabbits in each group were found not to be pregnant. Despite these problems, at least 10 litters from each dose group were available for examination.

Dams in each dose group showed better weight gains than controls, but there was no clear dose relationship. There were no adverse effects on the incidences of resorptions, numbers of live fetuses, fetal weights or the incidences of malformations. However, there were increased incidences of delayed ossification, suggestive of fetotoxicity, in the 100 and 300 µg/kg bw per day groups. The NOEL was 30 µg/kg bw per day (Lehman, 1969b).

Groups of 10 female Russian/Biberach rabbits were given daily oral doses of 0, 0.01, 1 or 50 mg/kg bw per day clenbuterol hydrochloride, in distilled water by gavage, on days 8 to 16 of gestation. There were no compound-related deaths, but body weight gain and food intake was reduced at the highest dose level.

At the highest dose level, the incidence of resorptions was significantly increased, and there was a corresponding decrease in the numbers of viable fetuses. Mean litter and fetal weights were significantly reduced at this level. There was an increased incidence of malformations, notably cleft palate and synostosis, in pups from dams given 50 mg/kg bw per day clenbuterol hydrochloride. The NOEL was 1 mg/kg bw per day (Kast, 1973c).

Groups of 13-16 mated New Zealand White rabbits were exposed to doses of clenbuterol hydrochloride calculated to be 0, 48, 146 or 300 µg/kg bw per day by the inhalation route using a fine bore tube sited over the tracheal opening. The heads of the animals were held in an air extraction chamber to minimize rebreathing of aerosol. Dosing took place on days 6-18 of gestation.

Signs of stress were observed in all groups as a result of the exposure procedures. Signs included anorexia, lethargy and respiratory distress, and some animals died in all groups, including the controls. There was a high incidence of malformations in the control groups which was probably related to the initial stress. There was no increased incidence in animals given 146 or 300 µg/kg bw per day.

As a result of the high numbers of malformations seen in control groups, no firm conclusions could be drawn from this study. However, the data suggested that doses of up to 300 µg/kg bw per day clenbuterol hydrochloride, when given by inhalation to pregnant rabbits on days 6-18 of gestation, were not teratogenic (Palmer et al., 1980).

Taken together, the animal data suggested that clenbuterol hydrochloride is teratogenic in rats and rabbits at relatively high doses. There was evidence of fetotoxicity. The NOEL for maternal toxicity and teratogenicity in the rat and rabbit following oral administration was 1 mg/kg bw per day. However, the NOEL for fetotoxicity in one study in the rabbit was 0.03 mg/kg bw per day.

2.2.6   Special studies on genotoxicity

Clenbuterol hydrochloride gave negative results in the Ames test with *Salmonella typhimurium* strains, in a reversion test with *Escherichia coli* strain WP2(P), in a gene mutation assay with Chinese hamster V79 cells, in an *in vivo* mouse micronucleus test and in an *in vivo* cytogenetic assay with Chinese hamster bone marrow.

In the mouse lymphoma test with L5178Y cells, no increased incidences in mutation frequency were observed in the absence of metabolic activation. However, with metabolic activation, one of two experiments gave positive results at the two highest concentrations used (Clements, 1992).

Viability counts were reduced at these concentrations and sampling error may have led to these spurious results (Kirkland, 1992).

In an *in vitro* assay with human lymphocytes, at concentrations of > 500 µg/ml, an increased incidence of chromosome aberrations was noted in the absence (but not in the presence) of metabolic activation. This was not dose-dependent.

The results are summarized in Table 3.

Table 3. **Results of genotoxicity studies on clenbuterol hydrochloride**

Test system	Test subject	Concentration	Results	Reference
Bacterial reversion	*Salmonella typhimurium* TA98, TA100, TA1535, TA1537, TA1538	40-2500 µg/plate	–	Baumeister, 1978
	Escherichia coli WP2(P)	40-1500 µg/plate	–	
Bacterial reversion	*S. typhimurium* TA98, TA100, TA1535, TA1537, TA1538	10-500 µg/plate	–	Ellenberger, 1985
Forward mutation assay	Chinese hamster V79 fibroblasts, HGPRT locus	10-100 µg/ml	–	Baumeister, 1985
Forward mutation assay	Mouse lymphoma L5178Y	300-800 µg/ml	–	Clements, 1992
In vitro cyto-genetics	human lymphocytes	177-2352 µg/ml	+/-	McEnaney, 1992
In vivo micro-nucleus test	mouse	0.006, 0.5, 5.0 mg/kg bw per day	–	Friedmann, 1982
In vivo cyto-genetics test	Chinese hamster (bone marrow)	19, 60, 186 mg/kg bw per day	–	Holmstrom & MacGregor, 1986

These data suggest that clenbuterol hydrochloride is not genotoxic.

2.2.7 Special studies on irritancy

2.2.7.1 Skin irritation

Clenbuterol hydrochloride was applied to the shaved skin of six Russian/Biberach rabbits, using an occlusive dressing, for a period of 28 days. No evidence of skin irritation was seen (Hewett & Notman, 1984). Similar results were noted in a further study of skin irritation (James, 1990d).

2.2.7.2 Intramuscular tolerance

Intramuscular injection of 0.5 ml of a formulation containing clenbuterol hydrochloride into the dorsal area of rabbits produced only minimal haemorrhage, oedema and necrosis (Kreuzer, 1988).

2.2.7.3 Eye irritation

A formulated commercial product containing clenbuterol hydrochloride (0.1 ml) was instilled into the left eye of six New Zealand White rabbits. The right eye served as a control. Mild irritation was noted and this resolved 48 hours after treatment (Bailey, 1990).

2.2.8 Special studies on immunotoxicity

Clenbuterol hydrochloride, as a formulated commercial product, was tested on the guinea-pig using the Buehler technique. The content of active ingredient was 70.4 μg/ml in the formulation. It was not a sensitizer in this study (James, 1990c).

Another study employed a 0.2% solution of clenbuterol hydrochloride in 30% aqueous ethanol utilizing guinea-pigs and the Magnusson and Kligman maximization test involving the use of Freund's adjuvant. No evidence of a sensitizing effect was found (Schuster, 1984).

2.2.9 Special studies on pharmacodynamic effects

2.2.9.1 Studies in animals

Clenbuterol is a β_2-sympathomimetic agent with a wide range of pharmacodynamic effects. It produces a potent, dose-dependent bronchiolytic effect, which was demonstrated in guinea-pigs following acetylcholine, histamine or bradykinin-induced bronchospasm. Similar effects have been reported in cats and dogs. Clenbuterol exerts positive chronotropic and inotropic effects on isolated atria. It induces tachycardia in rats, dogs, cats and a variety of farm animals, accompanied by reductions in systolic and diastolic blood pressure. In the isolated uterine horn of the estrus rat, it exerts a pronounced relaxing effect on serotonin, oxytocin and bradykinin-induced spasm. It has also been shown to exert a relaxing effect on smooth muscle in the guinea-pig ileum and to reduce intestinal mobility in the rat and mouse.

Clenbuterol results in a glycogenolytic effect in the rat myocardium associated with β-stimulation. It has no effect on hepatic glycogen. It causes a potent dose-dependent reduction of the gastric secretion.

In mice, it causes decreases in spontaneous activity and also leads to increases in barbiturate-induced sleeping time. The drug results in dose-dependent striated muscle relaxation. However, it has no antipyretic or analgesic effects in mice (Engelhardt, A., 1971; Engelhardt, G., 1971, 1976).

Clenbuterol exerted major relaxing effects on spasms induced by carbachol in guinea-pig trachea and lung. The effects were antagonised by propranolol (Landry, 1983). It also led to decreases in perfusion pressure of hind limb blood vessels, inhibitions of contractions of the uterus, increases in atrial rate and inhibition of electrically stimulated contractions of the ileum (O'Donnell, 1976). It improved the rates of mucociliary clearance in the guinea-pig (Streller, 1981).

In beagle dogs, oral doses of clenbuterol hydrochloride at 1.5, 3.0, 7.5 and 15.0 μg/kg bw resulted in a dose-dependent tachycardia at $\geq$ 3.0 μg/kg bw. There was a slight tachyphylactic effect demonstrated by a diminution of the intensity and duration of the response. There was a dose-related reduction in blood pressure and the changes in diastolic and systolic pressures ran in parallel. These changes occurred at the lowest dose used (Ueberberg, 1976).

In cattle, intravenous doses of 0.6 μg/kg bw and 1.5 μg/kg bw produced no cardiac effects, as shown by ECG examination, whereas 2.85 μg/kg bw produced pronounced tachycardia. A dose of 0.3 mg/kg bw clenbuterol resulted in tocolysis lasting approximately 5 hours (Ballarini, 1978). Different lag times in parturition may be induced depending on the dose and route of treatment; no adverse effects on the dam or calf were noted (Arbeiter & Thurner, 1977).

Four metabolites of clenbuterol that had been shown to be present in the kidneys and liver of treated target animals were tested for pharmacodynamic activity. Of these, only one, N-A 1141 (Figure 1) was shown to have activity in the guinea-pig. Its bronchiolytic effect was less than 20% that of clenbuterol, and, moreover, it accounted for only 12% of residues in the liver and kidney 6 hours after treatment (Engelhardt, 1971).

2.2.9.2   Studies in humans

Patients given 10 μg (0.167 μg/kg bw) clenbuterol by the inhalation route showed no signs of tachycardia as determined by ECG. There were slight decreases in blood pressure. When patients with cardiac arrhythmia were given this dose of clenbuterol, none of the patients showed any changes attributable to the drug (Hufnagel, 1974).

After inhalation exposure of patients with airways disease, onset of broncholytic action occurred within 5 minutes and lasted for up to 6 hours. Pulmonary function effects were greatest with doses of $\geq$ 5 μg (0.083 μg/kg bw); only a minimal effect was seen with 2.5 μg (0.042 μg/kg bw) (Schuster et al., 1989).

2.3   Observations in humans

A single blinded cross-over study was carried out to examine the acute bronchospasmolytic effect and possible side-effects following oral administration of placebo and three doses of clenbuterol (1, 2.5 and 5 μg/day) in patients (three male and three female) with chronic obstructive airway disease over a 3 day period. The drug was given orally, diluted with water. Observations were carried out over a 2-hour period following dosing. The average age of the patients was 55.7 years and they had an average body weight of 73.16 kg.

None of the 3 doses produced any clear, consistent effects on bronchial resistance, thoracic gas volume, radial pulse frequency or blood pressure, and no side effects were seen.

The pharmacological NOEL in this study was 5 μg/day, equivalent to 0.08 μg/kg bw per day (Kaik, 1978).

The bronchospasmolytic effect was examined in two groups of patients:

Group A: ten patients aged 46-75 years with chronic obstructive respiratory disease resulting from pulmonary tuberculosis.

Group B: five patients aged 56-67 years with chronic obstructive respiratory disease not related to tuberculosis, plus five patients aged 34-57 with bronchial asthma.

The bronchospasmolytic effect was examined after single oral doses of 1, 2.5, 5, 10, 20, 25 or 30 μg/person, and after a placebo dose.

In Group A patients, intrathoracic gas volume was significantly reduced and vital capacity and pneumometer values significantly increased at all dose levels. In Group B patients, airway resistance was significantly reduced, but no dose relationship could be demonstrated. No significant placebo effect was seen in either group. When compared with placebo values, a significantly greater increase for both vital capacity and pneumometer values was observed in Group A, even at the lowest dose used in this group (5 μg). However, at the two lowest doses used in Group B (1 and 2.5 μg), there were no significant differences from placebo values. The pharmacological NOEL in this study was 2.5 μg, equivalent to 0.042 μg/kg bw (Nolte & Laumen, 1972; Nolte, 1980).

Children who consumed between 0.05-0.075 mg of clenbuterol showed only mild tachycardia. A 30-year-old woman who consumed 30 tablets equivalent to 0.6 mg clenbuterol (10 μg/kg approximately) developed tachycardia and slight hypertension approximately 1 hour after consumption. No tablet remains were found on gastric lavage, and medicinal charcoal and a saline laxative were given. The following day, the patient's pulse rate and blood pressure had returned to normal (Boehringer, 1991a).

Patients (100+) administered doses of 20-60 μg/day (0.3-1.0 μg/kg bw per day) for up to 1 year or 20 μg/day (0.3 μg/kg bw per day) for up to 6 months showed no adverse effects except for slight tremor and occasional, mild tachycardia (Laumen, 1978; Tullgren & Lins, 1987).

3.   COMMENTS

The Committee considered toxicological data on clenbuterol, including the results of acute, short-term and reproductive toxicity studies, as well as studies on teratogenicity, genotoxicity and carcinogenicity. Results of pharmacokinetic and pharmacodynamic studies in animals and humans were also considered.

Clenbuterol is well absorbed after oral administration in a number of animal species and in humans. An oral dose is largely and rapidly excreted in the urine, and the majority of the remainder is excreted in the faeces. The biotransformation of clenbuterol is complex and a number of metabolites are formed. The major compound found in a number of species was unchanged clenbuterol. After oral administration of therapeutic doses to lactating cattle, clenbuterol was found in the milk.

When radiolabelled clenbuterol was given orally to pregnant rats, dogs, baboons and cattle, radioactivity was detected in the fetuses.

Clenbuterol was moderately toxic in mice and rats after oral administration, LD_{50} values being in the range of 80-175 mg/kg bw. It was less toxic in the dog (LD_{50} = 400-800 mg/kg bw). It was more toxic after parenteral administration, with LD_{50} values in the range of 30-85 mg/kg bw after intravenous administration. The main signs of toxicity included lethargy, tachycardia and tonic-clonic convulsions after oral administration.

The main effects noted in the repeat-dose studies were tachycardia and, at higher doses, myocardial necrosis. These effects are common with β-agonist drugs. The myocardial necrosis was considered to be secondary to hypoxia, due to reduced myocardial perfusion at a time of high oxygen demand resulting from increased cardiac rate.

In 30-day repeat-dose studies in mice and rats, NOELs of 2.5 and 1 mg/kg bw per day, respectively, were identified, largely based on cardiac lesions. However, in a range of repeat-dose studies in rats using doses of 0.01 to 100 mg/kg bw per day for durations of up to 18 months, administered through the oral, intravenous and inhalation routes, no NOELs were identified. Effects were usually related to cardiac function and were seen even at the lowest doses used. Similarly, no NOELs could be identified in a range of repeat-dose oral studies in dogs. These studies used doses ranging from 0.1 to 40 mg/kg bw per day. In a 26-week inhalation study in cynomolgus monkeys, the NOEL was 25 μg/kg bw per day, based on a number of observations including cardiac effects.

No evidence of carcinogenicity was noted in a two-year oral study in mice with doses of up to 25 mg/kg bw per day. In a two-year study with doses of up to 25 mg/kg bw per day in the Chbb:THOM rat, no evidence of carcinogenicity was seen. However, in Sprague-Dawley rats given 25 mg clenbuterol/kg bw per day orally for 2 years, an increased incidence of mesovarian leiomyomas occurred. With the related compounds salbutamol in rats and medroxalol in mice, the effects could be abolished by administration of the β-blocking agent propranolol. Mesovarian leiomyomas in rats and uterine leiomyomas in mice are known to occur following long-term treatment with β-adrenoceptor agonists and the Committee concluded that these were due to adrenergic stimulation and not to any genotoxic mechanism. Clenbuterol was not genotoxic in a range of *in vitro* and *in vivo* genotoxicity studies.

Epidemiological studies indicate that there have been no increased incidences of uterine leiomyomas in women following the use of β-adrenoceptor agonists.

Clenbuterol had no effects on fertility in a reproductive toxicity study in rats using oral doses of 1-50 mg/kg bw per day from 10 weeks prior to mating in males and two weeks prior to mating in females. However, doses of 50 mg/kg bw per day resulted in the deaths of pups soon after birth. To investigate the cause of the high pup mortality at this dose level, the litters of control dams were exchanged with those from dams given 50 mg/kg bw per day. Pups from rats given 50 mg/kg bw per day died on the first day of lactation regardless of whether they suckled on treated or control dams. The mechanism involved in this lethal effect is unknown. A NOEL was not identified in this study, because pup weights at birth were reduced in all treated animals.

In a reproductive toxicity study in which male rats were treated with 1.5-15 μg clenbuterol/kg bw per day orally for 70 days prior to mating and females with the same dose range for 14 days prior to mating, no adverse effects on reproduction were noted. The NOEL was 15 μg/kg bw per day.

In teratogenicity studies in rats, oral doses of 10 and 100 mg/kg bw per day produced teratogenic effects that included hydrocephalus, anasarca, umbilical hernia, anophthalmia, rib variations and splintering of vertebrae. These were accompanied by signs of maternal toxicity. The NOEL was 1 mg/kg bw per day. In three studies in rabbits using doses of 30 μg to 50 mg per kg bw per day, signs of fetotoxicity, including delayed ossification and cleft palate, occurred. The NOEL was 30 μg/kg bw per day.

Clenbuterol produced a range of pharmacodynamic effects in a number of animal species including tachycardia, hypertension and muscle relaxing effects. These were seen at single doses as low as 0.8 μg/kg bw.

Four metabolites of clenbuterol that had been shown to be present in the kidneys of treated target animals were tested for pharmacological activity. Of these, only one (N-A 1141) was shown to have activity. Its broncholytic effect in the guinea-pig was less than 20% that of clenbuterol. In addition, it accounted for only 1-2% of residues in the liver and kidney of target animals 6 hours after treatment.

In humans, clenbuterol produced a bronchiolytic effect when a single dose of 10 μg (0.167 μg/kg bw) was given by the inhalation route, but no evidence of tachycardia was seen at this dose. With oral doses of clenbuterol of up to 5 μg/day (0.08 μg/kg bw per day) over a 3-day period, there were no effects on bronchial resistance, thoracic gas volume, cardiac rate or blood pressure. The NOEL in this study was 5 μg/day (0.08 μg/kg bw per day). In a study to investigate the bronchospasmolytic effect in humans, patients with obstructive lung disease were given oral doses of up to 30 μg per person. Patients administered doses of 5 μg or more exhibited bronchospasmolytic effects, and the pharmacological NOEL in this study was 2.5 μg per person, equivalent to 0.04 μg/kg bw.

4.   EVALUATION

The Committee considered the most relevant study for determining the ADI to be that concerning the bronchospasmolytic effect in humans. The patients had chronic obstructive airway disease and thus were likely to be a very sensitive population for this effect. The NOEL identified in this study (2.5 μg per person, equivalent to 0.04 μg/kg bw) is approximately 25% of the dose in another study in which the inhalation route was used, but in which cardiac effects were not observed. This NOEL is approximately 50% of the oral dose used in another study where, again, cardiac effects did not occur. Hence, this NOEL for the bronchospasmolytic effect offers an additional safety margin for cardiac effects. The Committee therefore established an ADI of 0-0.004 μg/kg bw, based on the NOEL of 0.04 μg/kg bw per day for pharmacodynamic effects in humans and a safety factor of 10.

5.   REFERENCES

Amemiya, K., Kudoh, M., Suzuki, H., Saga, K., & Hosaka, K. (1984). Toxicology of mabuterol. *Arzneim.forsch.*, **34**(11a), 1680-1684.

Arbeiter, K. & Thurner, M. (1977). Effect of the sympathomimetic Planipart (NAB 365) on parturition in cattle. *Tierärztl. Umsch.*, **32**, 423-427.

Bailey, D.E. (1990). Primary eye irritation study in rabbits with Ventipulmin. Unpublished report No. TX-9005/Ventipulmin from Hazleton Laboratories America, Inc., Vienna, VA, USA. Submitted to WHO by Boehringer Ingelheim Vetmedica GmbH, Ingelheim am Rhein, Germany.

Baillie, H.W., Cameron, B.D., Draffan, G.H., & Schmid, J. (1980). Investigations of the placental transfer of ^{14}C-N-AB 365 CL in the cow. Unpublished report No. 111674 from Inveresk Research International. Submitted to WHO by Boehringer Ingelheim GmbH, Ingelheim am Rhein, Germany.

Balazs, T. & Bloom, S. (1982). Cardiotoxicity of adrenergic and vasodilating antihypertensive drugs. In: van Stee, E.W. (ed.), Cardiovascular Toxicology, Raven Press, New York, pp. 199-220.

Ballarini, G. (1978). Initial studies of the clinical pharmacology of the β_2 sympathomimetic agent NAB 365 (Planipart) and its use for pharmacological tocolysis in cows. *Tierärztl. Umsch.*, **33**, 421-427.

Baumeister, M. (1978). Mutagenicity studies with the substance N-AB 365 CL in the plate incorporation assay (Ames test). Unpublished report. Submitted to WHO by Boehringer Ingelheim Vetmedica GmbH, Ingelheim am Rhein, Germany.

Baumeister, M. (1985). Mutagenicity study with the substance N-AB 365 CL in the V79 (HGPRT)-test. Unpublished report. Submitted to WHO by Boehringer Ingelheim Vetmedica GmbH, Ingelheim am Rhein, Germany.

Boehringer (1991a). Summary of case reports of overdosage with clenbuterol. Unpublished report. Submitted to WHO by Boehringer Ingelheim Vetmedica GmbH, Ingelheim am Rhein, Germany.

Boehringer (1991b). Estimated transdermal absorption of clenbuterol hydrochloride following skin contact with granules and injection solutions for veterinary use. Unpublished report. Submitted to WHO by Boehringer Ingelheim Vetmedica GmbH, Ingelheim am Rhein, Germany.

Cameron, B.D. & Phillips, M.W.A. (1987a). The residue kinetics of [^{14}C]-NAB 365 CL in the calf. Unpublished report No. 4371 from Inveresk Research International. Submitted to WHO by Boehringer Ingelheim Vetmedica GmbH, Ingelheim am Rhein, Germany.

Cameron, B.D. & Phillips, M.W.A. (1987b). The .residue kinetics of (^{14}C)-NAB 365 CL in the lactating cow. Unpublished report No. 4445 from Inveresk Research International. Submitted to Boehringer Ingelheim Vetmedica GmbH, Ingelheim am Rhein, Germany.

Cameron, D.M., Crook, D., & Brown, G. (1992). Ventipulmin injection solution and Ventipulmin granules. Target species tolerance study in calves. Unpublished report No. BOI 139/921030 from Huntingdon Research Centre. Submitted to WHO by Boehringer Ingelheim Vetmedica GmbH, Ingelheim am Rhein, Germany.

Chasseaud, L.F., Cresswell D.G., & Savage J.A. (1978). Pharmacokinetics of ^{14}C-N-AB 365 during chronic inhalational toxicity in cynomolgus monkeys. Unpublished report No. U79-0212 from Huntingdon Research Centre. Submitted to WHO by Boehringer Ingelheim Vetmedica GmbH, Ingelheim am Rhein, Germany.

Clark, G.C., Collins, C.J., Heywood, R., Street, A.E., Gibson, W.A., Prentice, D.E., Edmonson, N., Lewis, D.J, & Mahjeed, S.K. (1979). Investigation of the effects on rats of inhalation of NAB 365 CL for 26 weeks. Unpublished report No. BOI 88/78768 from Huntingdon Research Centre. Submitted to WHO by Boehringer Ingelheim Vetmedica GmbH, Ingelheim am Rhein, Germany.

Clements, J. (1992). Study to determine the ability of clenbuterol hydrochloride to induce mutations at the thymidine kinase (tk) locus in mouse lymphoma L5178Y cells using a fluctuation assay. Unpublished report No. 2TKREBSG.002 from Hazleton Microtest. Submitted to WHO by Boehringer Ingelheim Vetmedica GmbH, Ingelheim am Rhein, Germany.

Collins, C.J., Clark, G.C., Heywood, R., Street, A.E., Hardy, C.J., Cherry, C.P., Gibson, W.A., Gopinath, C., Harrington S.M., Edmonson, N.A., & Howard, E. (1979). Investigation of the effects on monkeys of inhalation of NAB 365 CL aerosol for a period of 26 weeks. Unpublished report No. BOI 90/78712 from Huntingdon Research Centre. Submitted to WHO by Boehringer Ingelheim Vetmedica GmbH, Ingelheim am Rhein, Germany.

De Busk, R.F. & Harrison, D.C. (1969). The clinical spectrum of papillary-muscle disease. *N. Engl. J. Med.,* **281**, 1458-1467.

Ellenberger, J. (1985). Clenbuterol (NAB-365-CL) testing for point-mutagenic activity with *Salmonella typhimurium.* Unpublished report. Submitted to WHO by Boehringer Ingelheim Vetmedica GmbH, Ingelheim am Rhein, Germany.

Engelhardt, A. (1971). Pharmacological exposé of the substance NAB365. Supplement to the exposé on the compound NAB 365. Supplementary investigations (May 1976) on the pharmacology of the substance NAB 365 and its metabolites. Unpublished report. Submitted to WHO by Boehringer Ingelheim Vetmedica GmbH, Ingelheim am Rhein, Germany.

Engelhardt, G. (1971). Results of a comparative pharmacological investigation of NAB 365 CL, NAB 365 CL-D and NAB365 CL. Unpublished report. Submitted to WHO by Boehringer Ingelheim Vetmedica GmbH, Ingelheim am Rhein, Germany.

Engelhardt, G. (1976). Pharmacological profile of the action of NAB 365 (clenbuterol), a new bronchodilator with a selective effect on the adrenergic β_2 receptors. *Arzneim. Forsch.,* **26**(7a), 1404-1420.

Erichsen, D.F. (1988). Dose titration of Ventipulmin syrup in horses with chronic obstructive pulmonary disease (COPD). Trial No: 37-295-86-3. Unpublished report. Submitted to WHO by Boehringer Ingelheim Vetmedica GmbH, Ingelheim am Rhein, Germany.

Erichsen, D.F. (1989). A determination of incremental dose tolerance of Ventipulmin syrup in the horse. Trial No.: Range finder 84-VI. Unpublished report. Submitted to WHO by Boehringer Ingelheim Vetmedica, Ingelheim am Rhein, Germany.

Fenner, A. (1982). Respiratory rate, minute volume and tidal volume in clinically healthy fattening cattle and those with bronchopneumonia - taking into account the effects of clenbuterol. Doctoral Thesis, Veterinary Faculty, Ludwig-Maximilians University of Munich. Submitted to WHO by Boehringer Ingelheim Vetmedica GmbH, Ingelheim am Rhein, Germany.

Friedmann, J.Ch. (1982). Study of the possible mutagenic activity of the substance "NAB365Cl" evaluated by the micronucleus test (oral route). Unpublished report. Submitted to WHO by Boehringer Ingelheim Vetmedica GmbH, Ingelheim am Rhein, Germany.

Gibson, J.P., Sells, D.M., Cheng, H.C., & Yuh, L. (1987). Induction of uterine leiomyomas in mice by medroxalol and prevention by propanolol. *Toxicol. Pathol.,* **15**, 468-473.

Gopinath, C. & Gibson, W.A. (1987). Mesovarian leiomyomas in the rat. *Environ. Health Perspect.,* **73**, 107-113.

Hamm, D. & Erichsen, D.F. (1989). Acute and subacute toxicity of Ventipulmin. Unpublished report. Submitted to WHO by Boehringer Ingelheim Vetmedica GmbH, Ingelheim am Rhein, Germany.

Hawkins, D.R., Elsom, L.F., de-Salis, C.M., Roberts, N.L., & Cameron, D.M. (1984). The disposition of the combination product ^{14}C-NAB 365 CL/trimethoprim/sulphadiazine after oral administration to horses. Unpublished report No. HRC/BOI 112/84912 from Huntingdon Research Centre. Submitted to WHO by Boehringer Ingelheim Vetmedica GmbH, Ingelheim am Rhein, Germany.

Hawkins, D.R., Elsom, L.F., de-Salis, C.M., Morris, G.R., Roberts, N.L., Cameron, D.M., Offer, J., & Fish, C. (1985a). The disposition of the combination product ^{14}C-NAB 365 CL/trimethoprim/sulphadiazine in calves. Unpublished report No. HRC/BOI 111/84680B from Huntingdon Research Centre. Submitted to WHO by Boehringer Ingelheim Vetmedica GmbH, Ingelheim am Rhein, Germany.

Hawkins, D.R., Elsom, L.F., de-Salis, C.M., Morris, G.R., Roberts, N.L., & Cameron, D.M. (1985b).The disposition of the combination product ^{14}C-NAB 365 CL/trimethoprim/sulphadiazine in dairy cows. Unpublished report No. HRC/BOI 111/84680A from Huntingdon Research Centre. Submitted to WHO by Boehringer Ingelheim Vetmedica GmbH, Ingelheim am Rhein, Germany.

Hawkins, D.R., Elsom, L.F., Dighton, M.H., Kaur, A., & Cameron, D.M. (1993a). The pharmacokinetics, metabolism and residues of ^{14}C-clenbuterol (^{14}C-N-AB 365 CL) following intramuscular administration to calves.

Unpublished report No. HRC/BOI 140/921418 from Huntingdon Research Centre. Submitted to WHO by Boehringer Ingelheim Vetmedica GmbH, Ingelheim am Rhein, Germany.

Hawkins, D.R., Elsom, L.F., & Dighton, M.H. (1993b). Investigation of the metabolic profiles in the livers of the horses following multiple oral administration. Unpublished report No. BOI 149/931490 from Huntingdon Research Centre. Submitted to WHO by Boehringer Ingelheim Vetmedica GmbH, Ingelheim am Rhein, Germany.

Hewett, Chr. & Notman J. (1984). Dermal tolerance test with repeated topical application in the rabbit. Unpublished report. Submitted to WHO by Boehringer Ingelheim Vetmedica GmbH, Ingelheim am Rhein, Germany.

Holmstrom, M. & MacGregor, D.B. (1986). NAB 365 CL. Cytogenetic study in Chinese hamsters. Unpublished report No 4085 from Inveresk Research International. Submitted to WHO by Boehringer Ingelheim Vetmedica GmbH, Ingelheim am Rhein, Germany.

Hufnagel, E. (1974). Results with a new broncholytic agent. Effect on heart rate, blood pressure and ECG. *Ärztl. Praxis,* **28**, 1350-1352.

Jack, D., Poynter, D., & Spurling, N.W. (1983). *Beta*-adrenoceptor stimulants and mesovarian leiomyomas in the rat. *Toxicology,* **27**, 315-320.

James, C.N. (1990a). Acute oral toxicity in rats with Ventipulmin syrup. Unpublished report No. TX-9003/VENTIPULMIN. Submitted to WHO by Boehringer Ingelheim Vetmedica GmbH, Ingelheim am Rhein, Germany.

James, C.N. (1990b). Acute dermal toxicity study in rabbits with Ventipulmin syrup. Unpublished report No. TX-9002/VENTIPULMIN. Submitted to WHO by Boehringer Ingelheim Vetmedica GmbH, Ingelheim am Rhein, Germany.

James, C.N. (1990c). Guinea pig sensitisation study - Buehler method using Ventipulmin syrup. Unpublished report No. Case #: 64551-05 from Arthur D. Little, Inc., Cambridge, MA, USA. Submitted to WHO by Boehringer Ingelheim Vetmedica GmbH, Ingelheim am Rhein, Germany.

James, C.N. (1990d). Primary dermal irritation study in rabbits with Ventipulmin syrup. Unpublished report TX-9001/Ventipulmin from Arthur D. Little, Inc., Cambridge, MA, USA. Submitted to WHO by Boehringer Ingelheim Vetmedica GmbH, Ingelheim am Rhein, Germany.

Johnston, A.M. & Dunsire, J.P. (1993). Plasma kinetics, metabolism, excretion and residue kinetics of [^{14}C]-N-AB 365 CL following oral administration to the horse. Unpublished report No. 8346 from Inveresk Research International. Submitted to WHO by Boehringer Ingelheim Vetmedica GmbH, Ingelheim am Rhein, Germany.

Johnston, A.M. & Jenner, W.N. (1976). Study of the metabolism of ^{14}C-labelled N-AB 365 in the male baboon. Unpublished report from Inveresk Research International. Submitted to WHO by Boehringer Ingelheim Vetmedica GmbH, Ingelheim am Rhein, Germany.

Johnston, W. (1987). Ventipulmin solution - Bovine safety trial. Unpublished report. Submitted to WHO by Boehringer Ingelheim Vetmedica GmbH, Ingelheim am Rhein, Germany.

Kaik, G. (1978). Clinico-pharmacological tests using the β_2-sympathomimetic NAB 365 (clenbuterol) in patients with obstructive respiratory disease. Unpublished report from the Medical University Clinic of Vienna. Submitted to WHO by Boehringer Ingelheim Vetmedica GmbH, Ingelheim am Rhein, Germany.

Kast, A. (1973a). Subacute toxicity with compound NAB 365 CL on mice, oral administration. Unpublished report. Submitted to WHO by Boehringer Ingelheim Vetmedica GmbH, Ingelheim am Rhein, Germany.

Kast, A. (1973b). Teratological testing with the compound N-AB 365-CL on rats, oral administration. Unpublished report. Submitted to WHO by Boehringer Ingelheim Vetmedica GmbH, Ingelheim am Rhein, Germany.

Kast, A. (1973c). Teratological testing with the compound NAB-365 CL on rabbits, oral administration. Unpublished report. Submitted to WHO by Boehringer Ingelheim Vetmedica GmbH, Ingelheim am Rhein, Germany.

Kast, A. (1975). Teratological testing with the compound N-AB 365 CL on rats, oral administration during the period of organogenesis. Unpublished report. Submitted to WHO by Boehringer Ingelheim Vetmedica GmbH, Ingelheim am Rhein, Germany.

Kast, A. & Tsunenari, Y. (1973a). Subacute toxicity with compound NAB 365 CL on rats, oral administration. Unpublished report. Submitted to WHO by Boehringer Ingelheim Vetmedica GmbH, Ingelheim am Rhein, Germany.

Kast, A. & Tsunenari, Y. (1973b). Chronic toxicity of compound NAB 365 CL on rats, oral administration per food. Unpublished report. Submitted to WHO by Boehringer Ingelheim Vetmedica GmbH, Ingelheim am Rhein, Germany.

Kast, A. & Tsunenari, Y. (1973c). Subacute toxicity with compound NAB 365 CL on rats, intravenous administration. Unpublished report. Submitted to WHO by Boehringer Ingelheim Vetmedica GmbH, Ingelheim am Rhein, Germany.

Kirkland, D.J. (1992). Review of clenbuterol hydrochloride mutagenicity data in relation to the proposed acceptable daily intake. Unpublished report. Submitted to WHO by Boehringer Ingelheim Vetmedica GmbH, Ingelheim am Rhein, Germany.

Koellmer, H., Stotzer, H., & Paulini, K. (1977). Subacute toxicity study of the substance N-AB 365-CL in rats using inhalation over a period of 13 weeks. Unpublished report. Submitted to WHO by Boehringer Ingelheim Vetmedica GmbH, Ingelheim am Rhein, Germany.

Kopitar, Z. (1969). Autoradiographic investigations on the distribution of [^{14}C]-N-AB 365 CL in rats and pregnant mice (ADME 1 B). Unpublished report No. U69-0108. Submitted to WHO by Boehringer Ingelheim Vetmedica GmbH, Ingelheim am Rhein, Germany.

Kopitar, Z. (1970). Pharmacokinetic investigations with [^{14}C]-N-AB 365 CL in rats (ADME 1 B). Unpublished report No. U69-0109. Submitted to WHO by Boehringer Ingelheim Vetmedica GmbH, Ingelheim am Rhein, Germany.

Kopitar, Z & Zimmer, A. (1973). Effect of repeated doses of NAB 365 CL on the pharmacokinetic profile in rats (ADME 1 D). Unpublished report No. U73-0158. Submitted to WHO by Boehringer Ingelheim GmbH, Ingelheim am Rhein, Germany.

Kreuzer, H. (1988). Intramuscular local tolerance in tests with Ventipulmin injectable solution *ad us. vet.* in rabbits. Unpublished report. Submitted to WHO by Boehringer Ingelheim Vetmedica GmbH, Ingelheim am Rhein, Germany.

Landry, Y. (1983). Review of the chief pharmacological properties, clinical structures and physiochemical properties of clenbuterol. Unpublished report from Laboratoire d'Allergopharmacologie, Strasbourg. Submitted to WHO by Boehringer Ingelheim Vetmedica GmbH, Ingelheim am Rhein, Germany.

Laumen, F. (1978). Investigation on long-term treatment and cumulation with clenbuterol. *Med. Mon.schr.*, **29**, 455-459.

Lehman, H. (1969a). Teratology investigations in pregnant rats with the substance N-AB 365 CL. Unpublished report. Submitted to WHO by Boehringer Ingelheim Vetmedica GmbH, Ingelheim am Rhein, Germany.

Lehman, H. (1969b). Teratology investigations in pregnant rabbits with the substance N-AB 365 CL. Unpublished report. Submitted to WHO by Boehringer Ingelheim Vetmedica GmbH, Ingelheim am Rhein, Germany.

Lehman, H. (1974). Testing of the compound NAB 365 CL for perinatal toxicity in rats. Unpublished report. Submitted to WHO by Boehringer Ingelheim Vetmedica GmbH, Ingelheim am Rhein, Germany.

Lehman, H. (1975). Study with substance N-AB 365 CL for fertility inhibiting, teratogenic and perinatal toxic effects in rats. Unpublished report. Submitted to WHO by Boehringer Ingelheim Vetmedica GmbH, Ingelheim am Rhein, Germany.

Lehman, H. (1980). Test of the substance N-AB 365 CL for fertility-inhibiting, embryotoxic and foetotoxic effects in rats. Unpublished report of study No. 78 D. Submitted to WHO by Boehringer Ingelheim Vetmedica GmbH, Ingelheim am Rhein, Germany.

Lehman, H. (1981). Test of the substance N-AB 365 CL for fertility-inhibiting, embryotoxic and foetotoxic effects in rats. Unpublished report of study No. 73 D. Submitted to WHO by Boehringer Ingelheim Vetmedica GmbH, Ingelheim am Rhein, Germany.

Leuschner, F., Leuschner, A., Schwerdtfeger, W., Otto, H., & Dontenwill, W. (1969). Report on subacute toxicity studies of NAB 365 p. o.- batch Z 6191 in beagle dogs. Unpublished report. Submitted to WHO by Boehringer Ingelheim Vetmedica GmbH, Ingelheim am Rhein, Germany.

McEnaney, S. (1992). Damaging potential of clenbuterol by its effects on cultured human lymphocytes using an *in vitro* cytogenetics assay. Unpublished report No. 2HLREBSG.002 from Hazleton Microtest. Submitted to WHO by Boehringer Ingelheim Vetmedica GmbH, Ingelheim am Rhein, Germany.

Magnusson, G. & Hansson, E. (1973). Myocardial necrosis in the rat: a comparison between isoprenaline, orciprenaline, salbutamol and terbutaline. *Cardiology,* **58**, 174-180.

Nelson, L.W. & Kelly, W.A. (1971). Mesovarian leiomyomas in rats in a chronic toxicity study of soterenol hydrochloride. *Vet. Pathol.,* **8**, 452-457.

Nishikawa, J., Tsunenari, Y., & Kast, A. (1972). Acute toxicity of NAB 365 CL in rats, mice and rabbits. Unpublished report. Submitted to WHO by Boehringer Ingelheim Vetmedica GmbH, Ingelheim am Rhein, Germany.

Nishikawa, J., Kast, A., & Klupp, H. (1979). Acute toxicity study with clenbuterol (NAB 365 CL) in rats dosed orally or intravenously. Unpublished report. Submitted to WHO by Boehringer Ingelheim Vetmedica GmbH, Ingelheim am Rhein, Germany.

Nolte, D. (1980). Comments on the publication "Lung function tests with the bronchospasmolytic agent NAB 365". Unpublished report. Submitted to WHO by Boehringer Ingelheim GmbH, Ingelheim am Rhein, Germany.

Nolte, D. & Laumen, F. (1972). Pulmonary function tests following the bronchospasmolytic agent NAB 365. Unpublished report from Medical Clinics and Polyclinics, Justus-Liebig University and Seltersberg Sanatorium of Hessen State Insurance Association, Giessen, Germany. Submitted to WHO by Boehringer Ingelheim Vetmedica GmbH, Ingelheim am Rhein, Germany.

O'Donnell, S.R. (1976). Selectivity of clenbuterol (NAB 365) in guinea-pig isolated tissues containing β-adrenoceptors. *Arch. Int. Pharmacodyn.*, **224**, 190-198.

Owen, R., Fischer, R., & Erichsen, D.F. (1990). Safety of Ventipulmin syrup in horses: 64-day study trial No. 37-295-83-4. Unpublished report. Submitted to WHO by Boehringer Ingelheim Vetmedica GmbH, Ingelheim am Rhein, Germany.

Palmer, A.K., Edwards, J.A., & Collins, C.J. (1978). Effect of NAB 365 CL metered aerosol on pregnancy of the rat. Unpublished report No. BOI 85/78689 from Huntingdon Research Centre. Submitted to WHO by Boehringer Ingelheim Vetmedica GmbH, Ingelheim am Rhein, Germany.

Palmer, A.K., Bottomley, A.M., & Collins, C.J. (1980). Effect of NAB 365 CL metered aerosol on pregnancy of the New Zealand White rabbit. Unpublished report No. BOI 86/79656 from Huntingdon Research Centre. Submitted to WHO by Boehringer Ingelheim Vetmedica GmbH, Ingelheim am Rhein, Germany.

Pappritz, G. (1968). Determination of ALD_{50} of the substance in the dog after oral administration. Unpublished report. Submitted to WHO by Boehringer Ingelheim Vetmedica GmbH, Ingelheim am Rhein, Germany.

Pappritz, G. (1972). Determination of the ALD_{50} 0f NA-B 365 in the dog following intravenous administration. Unpublished report. Submitted to WHO by Boehringer Ingelheim Vetmedica GmbH, Ingelheim am Rhein, Germany.

Pappritz, G. & Bauer, M. (1973a). Supplementary studies on the subacute toxicity of substance N-AB 365 CL in dogs after oral administration. Unpublished report. Submitted to WHO by Boehringer Ingelheim Vetmedica GmbH, Ingelheim am Rhein, Germany.

Pappritz, G. & Bauer, M. (1973b). Chronic toxicity study of substance NAB 365 in dogs following oral administration. Unpublished report. Submitted to WHO by Boehringer Ingelheim Vetmedica, Ingelheim am Rhein, Germany.

Pappritz, G., Bauer, M., & Eckenfels, A. (1973). Subacute toxicity of NAB 265 CL in dogs following intravenous administration. Unpublished report. Submitted to WHO by Boehringer Ingelheim Vetmedica, Ingelheim am Rhein, Germany.

Poynter, D & Spurling, N.W. (1971). Some cardiac effects of beta-adrenergic stimulants in animals. *Postgrad. Med. J.,* **47**, 21-25.

Poynter, D., Harris, D.M., & Jack, D. (1978). Salbutamol: lack of evidence of tumour induction in man. *Br. Med. J.,* **7 January**.

Richter, I. (1982). Biochemical investigations of the placental passage of ^{14}C-clenbuterol in pregnant rats. Unpublished report No. U82-0292. Submitted to WHO by Boehringer Ingelheim Vetmedica GmbH, Ingelheim am Rhein, Germany.

Roberts, W. C., & Cohen, L.S. (1972). Left ventricular papillary muscles. Description of the normal and a survey of conditions causing them to be abnormal. *Circulation,* **XLVI**, 138-154.

Rohr, U.D., Haczkiewicz, K., Mohr, A. , & Orton, B. (undated). Validation of an excised human skin model for testing drug candidates suited for transdermal drug delivery. Unpublished report from Boehringer Ingelheim KG. Submitted to WHO by Boehringer Ingelheim Vetmedica GmbH, Ingelheim am Rhein, Germany.

Rominger, K.L. & Schrank, H. (1982). Placental transfer of ^{14}C-clenbuterol in the dog. Unpublished report No. U82-0291. Submitted to WHO by Boehringer Ingelheim Vetmedica GmbH, Ingelheim am Rhein, Germany.

Rominger, K.L., Förster, H., Hermer, M., Peil, H., & Wolf, M. (1987). Clenbuterol plasma levels in patients under tocolytic treatment. Unpublished report from Boehringer Ingelheim KG. Submitted to WHO by Boehringer Ingelheim Vetmedica GmbH, Ingelheim am Rhein, Germany.

Rosenblum, I., Wohl, A., & Stein, A.A. (1965). Studies in cardiac necrosis. III. Metabolic effects of sympathomimetic amines producing cardiac lesions. *Toxicol. Appl. Pharmacol.,* 7, 344-351.

Schmid, J. (1977). Pilot pharmacokinetic investigations after intravenous administration of N-AB 365 CL in a cow. Unpublished report. Submitted to WHO by Boehringer Ingelheim Vetmedica GmbH, Ingelheim am Rhein, Germany.

Schmid, J. (1980). Investigations of the placental transfer of ^{14}C-N-AB 365 CL in the baboon. Unpublished report from Inveresk Research International. Submitted to WHO by Boehringer Ingelheim Vetmedica GmbH, Ingelheim am Rhein, Germany.

Schmid, J. (1982). Pharmacokinetics, metabolism and tissue distribution of ^{14}C-N-AB 365 CL in the baboon. Unpublished report No. 2220 from Inveresk Research International. Submitted to WHO by Boehringer Ingelheim Vetmedica GmbH, Ingelheim am Rhein, Germany.

Schmid, J. (1990a). Plasma levels and residue analysis in cows and calves. Unpublished report No. U90-0092. Submitted to WHO by Boehringer Ingelheim Vetmedica GmbH, Ingelheim am Rhein, Germany.

Schmid, J. (1990b). N-AB 365 CL in the cow's milk after oral and intramuscular administration. Unpublished report No. U90-0099. Submitted to WHO by Boehringer Ingelheim Vetmedica GmbH, Ingelheim am Rhein, Germany.

Schmid, J & Prox, A. (1986). Isolation and structural elucidation of NAB 365 CL metabolites from dog urine. Unpublished report. Submitted to WHO by Boehringer Ingelheim Vetmedica GmbH, Ingelheim am Rhein, Germany.

Schmid, J. & Zimmer, A. (1977a). Pilot pharmacokinetic studies following oral administration (single and multiple dosing) of ^{14}C N-AB 365 CL to a cow. Unpublished report. Submitted to WHO by Boehringer Ingelheim Vetmedica GmbH, Ingelheim am Rhein, Germany.

Schmid, J. & Zimmer, A. (1977b). Pilot pharmacokinetic investigations after intramuscular administration (single and multiple doses) of N-AB 365 CL in the cow. Unpublished report. Submitted to WHO by Boehringer Ingelheim Vetmedica GmbH, Ingelheim am Rhein, Germany.

Schuster, A. (1984). Skin sensitisation study in guinea pigs. Unpublished report No. U84-0673 from Boehringer Ingelheim KG. Submitted to WHO by Boehringer Igelheim Vetmedica GmbH, Ingelheim am Rhein, Germany.

Schuster, D., Deichsel, G., & Feuerer, W. (1989). Dose effect study with NAB 365 CL metered dose inhaler in patients with chronic obstructive lung disease. Unpublished report from Dr. K. Thomae GmbH, Medical Department. Submitted to WHO by Boehringer Ingelheim Vetmedica GmbH, Ingelheim am Rhein, Germany.

Sells, D.M. & Gibson, J.P. (1987). Carcinogenicity studies with medroxalol hydrochloride in rats and mice. *Toxicol. Pathol.*, 15, 457-467.

Serbedija, R. & Bauer, M. (1973). Toxicity study of the substance NAB 365 CL in rats over a period of 18 months. Unpublished report. Submitted to WHO by Boehringer Ingelheim Vetmedica GmbH, Ingelheim am Rhein, Germany.

Serbedija, R., Lutzen, L., Puschner, H., & Hohbach, Ch. (1982). Carcinogenicity study with the substance N-AB 365 CL in rats with oral administration for a period of 2 years. Unpublished report. Submitted to WHO by Boehringer Ingelheim Vetmedica GmbH, Ingelheim am Rhein, Germany.

Streller, I. (1981). Effect of the β-adrenergic agents fenoterol, orciprenaline, clenbuterol, SOM 987 and SOM 1122 on the mucocilliary clearance in the trachea of the anaesthetised guinea-pig. Unpublished report. Submitted to WHO by Boehringer Ingelheim Vetmedica GmbH, Ingelheim am Rhein, Germany.

Tullgren, A & Lins L-E. (1987). An open long-term tolerability study on NAB 365 (clenbuterol). Unpublished report from Boehringer Ingelheim AB, Stockholm, Sweden. Submitted to WHO by Boehringer Ingelheim Vetmedica GmbH, Ingelheim am Rhein, Germany.

Ueberberg, H. (1976). Comparative study of the effects of NAB 365 CL and terbutaline sulfate on the dog heart. Unpublished report. Submitted to WHO by Boehringer Ingelheim Vetmedica GmbH, Ingelheim am Rhein, Germany.

Umemoto, K. (1984). Carcinogenicity studies with the *beta*-adrenoceptor stimulant clenbuterol (NAB-365 CL) in mice by drinking water. Unpublished report from Nippon Boehringer Ingelheim. Submitted to WHO by Boehringer Ingelheim Vetmedica GmbH, Ingelheim am Rhein, Germany.

Zimmer, A. (1971). Metabolism and species comparison in the rat, rabbit and dog (ADME IV). Unpublished report. Submitted to WHO by Boehringer Ingelheim Vetmedica GmbH, Ingelheim am Rhein, Germany.

Zimmer, A. (1974a). Comparison of the pharmacokinetic profile in the dog with single and repeated dosage (ADME 1 D). Unpublished report No. U73-0161. Submitted to WHO by Boehringer Ingelheim GmbH, Ingelheim am Rhein, Germany.

Zimmer, A. (1974b). Pharmacokinetics and metabolism pattern in the rabbit and dog (ADME I). Unpublished report No. U74-0116. Submitted to WHO by Boehringer Ingelheim GmbH, Ingelheim am Rhein, Germany.

Zimmer, A. (1974c). Metabolism in man; comparison with dog and rabbit, using radioactive substance. Unpublished report. Submitted to WHO by Boehringer Ingelheim Vetmedica GmbH, Ingelheim am Rhein, Germany.

Zimmer, A. (1976). Administration of clenbuterol in man. Single doses, multiple doses, and metabolite samples. Unpublished report. Submitted to WHO by Boehringer Ingelheim Vetmedica GmbH, Ingelheim am Rhein, Germany.

Zimmer, A. (1977). Preliminary investigations of pharmacokinetics in a horse. Unpublished report. Submitted to WHO by Boehringer Ingelheim Vetmedica GmbH, Ingelheim am Rhein, Germany.

Zupan, J. A. (1985). A comparative bioavailability study between oral and injectable dosage forms of clenbuterol in horses. Unpublished report. Submitted to WHO by Boehringer Ingelheim Vetmedica GmbH, Ingelheim am Rhein, Germany.

XYLAZINE

First draft prepared by
Dr Pamela L. Chamberlain
Center for Veterinary Medicine
Food and Drug Administration,
Rockville, Maryland, USA

1.   EXPLANATION

Xylazine is a clonidine analogue that acts on presynaptic and postsynaptic receptors of the central and peripheral nervous systems. It is an α_2-adrenergic agonist used in animals, including cattle, horses, dogs, cats and deer, for its tranquillizing, muscle relaxant and analgesic effects, but it has numerous other pharmacological effects. It inhibits the effects of postganglionic cholinergic nerve stimulation.

Xylazine is administered by the intramuscular, intravenous or subcutaneous (in cats) routes, often in combination with other anaesthetic agents, e.g., barbiturates, chloral hydrate, halothane and ketamine.

Xylazine had not been previously evaluated by the Committee. The molecular structure of xylazine is shown below.

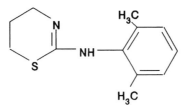

2. BIOLOGICAL DATA

2.1 Biochemical aspects

2.1.1 Pharmacodynamics

With respect to xylazine's sedative effect, there are marked species differences in the dose rates required to achieve this state. Table 1 illustrates dosages required for various animal species (Gross & Tranquilli, 1989).

Table 1. Dosage of xylazine in various animal species

| Species | Xylazine (mg/kg bw) | |
	Intravenous	Intramuscular
Horses	0.5 to 1.1	1 to 2
Cattle	0.03 to 0.1[1]	0.1 to 0.2[1]
Sheep	0.05 to 0.1[1]	0.1 to 0.3[1]
Goats	0.01 to 0.5[1]	0.05 to 0.5[1]
Swine		2 to 3
Dogs	0.5 to 1	1 to 2
Cats	0.5 to 1	1 to 2
Birds		5 to 10

[1] Lower end of dose range should be used if sedation without recumbency is desired (Gross & Tranquilli, 1989).

Mydriasis is a feature of xylazine-induced sedation in the cat. The mechanism has been determined as central inhibition of parasympathetic tone in the iris due to xylazine's activation of post-synaptic alpha-2 receptors (Hsu *et al.*, 1981).

Thermoregulatory control is impaired in cats administered xylazine. They become more susceptible to hyper- and hypothermia both during and after recovery from the sedative effects of the drug. Foals have demonstrated a hypothermic response to xylazine. Thermoregulatory effects in cattle have been variable (Ponder & Clark, 1980; Booth, 1988; Robertson *et al.*, 1990).

Cardiovascular effects of xylazine include decreased heart rate and variable effects on blood pressure. Xylazine-induced arrhythmia is common in the horse due to sinoatrial and atrioventricular blocks. Arrhythmias have also been recorded in dogs, but could not be induced in sheep. The induction of cardiovascular effects may be influenced by route of administration, e.g., xylazine administered epidurally to horses produced no cardiovascular changes, whereas cattle injected by this route experienced decreases in heart rate and arterial blood pressure (Sagner *et al.*, 1969; Holmes & Clark, 1977; Freire *et al.*, 1981; Hsu *et al.*, 1981; Wasak, 1983; Singh *et al.*, 1983; Leblanc & Eberhart, 1990; Skarda *et al.*, 1990).

The effects of xylazine on respiration, acid-base balance and blood gas values vary according to species and anaesthetic combination. In cattle, xylazine causes a slowing of the respiratory rate. This is accompanied by an increase in pH and metabolic acidosis. Respiratory rate is also slowed in dogs administered xylazine, but arterial pH, pO_2 or pCO_2 are not significantly affected. The literature contains conflicting reports on the effect of xylazine on the respiratory rate of horses. Tachypnoea is characteristic of the ovine response to xylazine. Hypoxaemia induced by xylazine in sheep can be life-threatening (DeMoor & Desmet, 1971; Klide *et al.*, 1975; Holmes & Clark, 1977; Hsu *et al.*, 1989; Carter *et al.*, 1990; Wagner *et al.*, 1991).

Hyperglycaemia is induced by xylazine in adults of all target species. Increased blood glucose concentrations are accompanied by a decrease in insulin levels. In adult horses, hyperglycaemia is accompanied by increased urine volume without glycosuria. Xylazine administered to neonatal foals did not result in hyperglycaemia. The hyperglycaemic effect of xylazine is thought to be due to its direct effect on alpha-2-adrenoceptors of pancreatic islet beta cells resulting in an inhibition of insulin release (Symonds, 1976; Feldberg & Symonds, 1980; Hsu & Hummel, 1981; Thurmon *et al.*, 1982, 1984; Benson *et al.*, 1984).

Serum chemistry and cerebral spinal fluid alterations were observed in adult female goats administered intramuscularly with 0.2 mg xylazine/kg bw. Significant elevations of urea nitrogen, total protein and total cholesterol were found in serum. Glucose and urea nitrogen levels were significantly increased (P<0.01) and chloride levels were significantly decreased (P<0.05) in the cerebral spinal fluid (Amer & Misk, 1980).

Erythrocyte counts, haematocrit values and haemoglobin concentrations in cattle and dogs have shown significant but reversible decreases following xylazine administration (Eichner et al., 1979; Wasak, 1983).

Gastrointestinal effects in ruminants include decreased gut motility, prolongation of gastrointestinal transit time and inhibition of reticulorumen contractions. Xylazine causes decreased muscle tone of the colon and rectum which facilitates rectal examination. Xylazine inhibition of rumen contractions can lead to tympany, which is a potential cause of death in xylazine-sedated ruminants. Ruminants are fasted prior to sedation and maintained in sternal recumbency during sedation to reduce the risk of xylazine-induced rumen tympany. Because xylazine also impairs deglutition, the head and neck of xylazine-sedated ruminants are lowered to avoid aspiration of saliva or ruminal fluid. Tolazoline (an alpha-2-adrenergic antagonist) has shown effectiveness in reversing recumbency, gastric paresis and loss of voluntary lingual control caused by xylazine in cattle (Swift, 1977; Bolte & Stupariu, 1978; Ruckebusch & Toutain, 1984).

Gastrointestinal effects in dogs and cats include decreased transit time and vomiting. The mechanism for induction of vomiting is thought to involve the effect of xylazine on alpha-2-adrenoceptors in the area postrema (the chemoreceptor trigger zone for vomiting) in the medulla oblongata (Cullen & Jones, 1977; Colby et al., 1981; Hsu & McNeel, 1983; Hikasa et al., 1987, 1989).

2.1.2 Absorption, distribution and excretion

2.1.2.1 Rats

Male Sprague-Dawley rats (170 g bw) were administered xylazine at dosages of 0.02 to 10 mg/kg bw (i.v.) or 0.02 to 100 mg/kg bw (oral). The drug was labelled with both ^{35}S and ^{14}C on the thiazine ring. Following oral administration, absorption was >95% with a half-life of approximately 5 minutes. After i.v. administration, the drug was distributed within a few minutes to almost all organs but primarily to the kidneys and central nervous system. Relatively high activity concentrations occurred in the pancreas, thyroid glands, liver and cranial glands (e.g., extraorbital, sublingual). Several hours following i.v. administration of 2 mg/kg bw, only small concentrations (< 0.3 μg/g tissue) were present in the musculature.

Following oral or i.v. administration, approximately 70% of the administered dose was eliminated in urine and 30% in faeces. Renal elimination following oral or i.v. administration was associated with a half-life of 2 to 3 hours. High oral doses (100 mg/kg bw) were associated with a delay in renal elimination. Faecal elimination was comparable to biliary elimination after oral or i.v. administration. Enterohepatic circulation did not occur to a notable extent (Duhm *et al.*, 1968, 1969).

2.1.2.2   Cattle

Three male calves (200-250 kg) and one dairy cow (450 kg) were injected intramuscularly with a 0.33 mg/kg dose of ^{14}C-xylazine labelled in the thiazine ring. Radioactivity in blood plasma reached its peak in the first 1.5 hours after injection. Total excretion of radioactivity in urine and faeces was 68, 86, 83 and 100% at 10, 24, 48 and 72 hours, respectively (Murphy & Jacobs, 1975).

In another study, five 2-month old calves and four lactating cows were administered a single intramuscular dose (0.3 or 0.6 mg/kg bw) of xylazine hydrochloride. Maximum concentrations of xylazine were achieved in blood 20 minutes after dosing. These were 0.04 mg/litre for the 0.3 mg/kg bw dose and 0.06 mg/litre for the 0.6 mg/kg bw dose. No xylazine was found in blood 8 hours after administration (Takase *et al.*, 1976).

Three lactating cows were administered an i.m. dose of 0.2 mg xylazine/kg bw and two others were administered an i.m. dose of 0.4 mg xylazine/kg bw. Milk was analysed for the presence of xylazine at 5 and 21 hours following administration. No xylazine was found at either time point for either dose. The limit of detection was 0.06 mg/litre (Pütter & Sagner, 1973).

Urinary excretion of xylazine was studied in three cows. Two were administered an i.m. dose of 0.2 mg xylazine/kg bw and one was administered an i.m. dose of 0.5 mg xylazine/kg bw. Less than 1% of the dose was excreted unchanged in the urine. Unchanged xylazine was no longer detectable 6 hours following administration. Metabolites were no longer detected in urine 10 hours after administration. The limit of detection for unchanged xylazine was 1-5 μg/litre (Pütter & Sagner, 1973).

2.1.2.3   Comparative pharmacokinetics in dogs, sheep, cattle and horses

The comparative pharmacokinetics of xylazine in dogs, sheep, cattle and horses are summarized in Table 2.

Table 2. Single-dose pharmacokinetics of xylazine in domestic species (Garcia-Villar *et al.*, 1981)

Species	Dog	Sheep	Cattle	Horse
Body weight range (kg)	14-24	42-65	240-440	415-550
Dose rate (mg/kg bw)[1]	1.4	1.0	0.2	0.6
Number	4	6	4	4
Intravenous[2]				
Distribution half-life (min)	2.57	1.89	1.21	5.97
Volume of distribution (l/kg)	2.52	2.74	1.94	2.46
Elimination half-life (min)	30.13	23.11	36.48	49.51
Body clearance (ml/min/kg)	81	83	42	21
Intramuscular[2]				
Absorption half-life (min)	3.44	5.45	ND	2.72
Elimination half-life (min)	34.65	22.36	ND	57.7
C_{max} (mg/ml)	0.43	0.13	ND	0.17
T_{max} (min)	12.7	14.68	ND	12.92
Bioavailability:				
mean (%)	73.9	40.8	ND	44.6
standard deviation (%)	17.89	23.81		4.16
range (%)	52-90	17-73		40-48

[1] Dosage expressed as xylazine-base
[2] Blood sampling times after injection: 1, 2, 4, 8, 16, 30 and 120 minutes.
ND = Not determined (assay was not sensitive enough to determine xylazine plasma concentrations lower than 0.01 mg/litre)

Pharmacokinetic parameters do not vary greatly between species following intravenous administration. The rapid elimination of xylazine is attributed to extensive metabolism, and not to rapid renal excretion of unchanged xylazine. Significant amounts of parent xylazine were not found in the urine of sheep collected at 10-minute intervals after dosing. The pharmacokinetics of xylazine were unmodified when it was administered to rabbits with occluded renal arteries. The lack of correlation between pharmacokinetic parameters and clinical effects of xylazine in cattle suggests that clinical effects in cattle are due to a rapidly produced long-acting metabolite(s) and not due to an increased sensitivity to xylazine (Garcia-Villar *et al.*, 1981).

2.1.3   Biotransformation

2.1.3.1   Rats

Studies were conducted with urine and bile of rats administered 2 mg xylazine (^{35}S or ^{14}C)/kg bw intravenously. Approximately 20 metabolites were detected and quantified as xylazine equivalents. Approximately 8% of the dose was eliminated as unchanged compound in the urine 24 hours after dosing. The major metabolite comprised 35% of the administered dose. Final products of metabolism were inorganic sulfate and carbon dioxide (Duhm et al., 1968).

Specific metabolites of xylazine were identified following incubation of xylazine with rat liver microsomes. Those metabolites were 2-(4'-hydroxy-2',6'-dimethylphenylamino)-5,6-dihydro-4H-1,3-thiazine, 2-(3'-hydroxy-2',6'-dimethylphenylamino)-5,6-dihydro-4H-1,3-thiazine, N-(2,6-dimethylphenyl)thiourea and 2-(2',6'-dimethylphenylamino)-4-oxo-5,6-dihydro-1,3-thiazine. N-(2,6-dimethylphenyl)thiourea was the major metabolite produced in vitro. Figure 1 shows the proposed metabolic pathways of xylazine based on these findings (Mutlib et al., 1992).

2.1.3.2   Horses

One mare was administered a 1 g dose of xylazine (route not stated) and urine was collected over 24 hours. Metabolites were recovered from horse urine only after the urine was hydrolysed with beta-glucuronidase. The major urinary metabolites detected were the same as those produced by incubating xylazine with rat liver microsomes, described in section 2.1.3.1 (Mutlib et al., 1992).

2.1.3.3   Cattle

Urine from three cows administered an i.m. dose of 0.2 mg xylazine/kg bw (two cows) or 0.5 mg xylazine/kg bw (one cow) was examined for metabolites. One urinary metabolite, identified as 2,6-xylidine[1], was found in both free and conjugated forms. The authors concluded that xylazine was essentially eliminated in cattle by rapid biotransformation. Breakdown of the thiazine ring, resulting in formation of 2,6-xylidine, was proposed as the primary biotransformation pathway (Pütter & Sagner, 1973).

---

[1]   2,6-xylidine is also known as 1-amino-2,6-dimethylbenzene and as 2,6-dimethylaniline

Figure 1. Proposed metabolic pathways of xylazine (from Mutlib *et al.*, 1992)

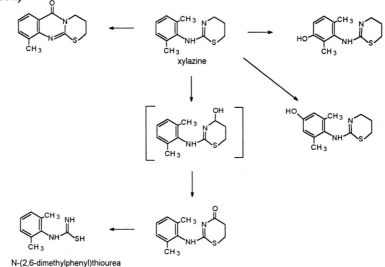

xylazine

N-(2,6-dimethylphenyl)thiourea

(structure in brackets was not isolated)

2.2   Toxicological studies

Because 2,6-xylidine is also a chemical intermediate used in dyes, a component of tobacco smoke and a degradation product of aniline-based pesticides, its toxicology has been studied extensively. Toxicological studies conducted with this compound were also reviewed and will be presented in addition to the review and results of toxicological studies of xylazine.

2.2.1   Acute toxicity studies of xylazine and 2,6-xylidine

The acute systemic toxicity of xylazine has been investigated in both laboratory and domestic species. It is generally recognized that ruminants are much more sensitive than most other species to the pharmacological and toxicological effects of xylazine.

Results of LD_{50} studies of xylazine and 2,6-xylidine are summarized in Table 3.

2.2.1.1   Acute toxicity of xylazine in dogs

Adult dogs (four males, four females) and cats (two males, four females) were administered a single i.m. or i.v. dose of 22 mg xylazine/kg bw (10 times the recommended therapeutic dose). One cat out of three receiving the i.v. dose died, and two dogs out of four receiving the i.m. dose died. All others recovered from convulsions, unconsciousness and respiratory depression with no apparent after-effects. The authors concluded that xylazine was slightly toxic in this study (Crawford *et al.*, 1970a).

Table 3. Results of acute toxicity studies on xylazine and 2,6-xylidine

Species	Sex[1]	Route	LD$_{50}$ (mg/kg bw)	Reference
Xylazine				
Rat	NA	p.o.	130	Sagner, 1967
Cat	male & female[2]	s.c.	100-110	Bauman & Nelson, 1969
Dog	male & female[3]	i.m	47	Nelson *et al.*, 1968b
Dog	4 male & 3 female	i.v	20-25	Nelson *et al.*, 1968b
Horses	NA[4]	i.m.	60-70[5]	Nelson *et al.*, 1968a
		i.v.	15-28[5]	
2,6-Xyladine				
Mouse	male	p.o.	710	Vernot *et al.*, 1977
Rat		p.o.	2042	Lindstrom *et al.*, 1969
	male	p.o.	840	Jacobson, 1972
	male	p.o.	630	Short *et al.*, 1983
	male	p.o.	1230	Vernot *et al.*, 1977
	female	p.o.	1160 & 1270	US National Toxicology Program, 1990
	male	p.o.	620-1250 & 1310	

[1] NA = Information not available
[2] Number per sex not stated; 10 animals were used
[3] Number per sex not stated; 17 animals were used
[4] Sex of test animals was not stated; 5 animals were used
[5] Minimum lethal dose

2.2.1.2 Acute toxicity of xylazine in horses

Adult horses were administered 11 mg xylazine/kg bw, i.v. (three mares, one gelding) or 22 mg xylazine/kg bw, i.m. (two mares, two geldings). One mare died following i.v. administration. All other test animals recovered from treatment-related effects 24 hours following i.v.

administration and 48 hours following i.m. administration. The authors concluded that the i.v. dose was slightly toxic and that the i.m. dose produced no apparent toxicity in this study (Crawford *et al.*, 1970b).

2.2.2 Short-term toxicity studies

2.2.2.1 Xylazine

a) Rats

Xylazine was administered in the diet to Wistar rats (10/sex/group) for 32 weeks. Dosages administered were 0, 50, 100, 250 or 500 mg/kg diet (equal to 0, 3, 6, 21 or 41 mg/kg bw per day for males and 0, 4, 8, 19 or 45 mg/kg bw per day for females). Haematology, urinalysis and gross and histopathological evaluations were performed.

Decreases in body weight observed in females in the two highest-dose groups (statistically significant ($p < 0.02$) at 500 mg/kg diet) were considered by the author to be treatment-related. Microscopic examination of livers, lungs and kidneys revealed that animals in all groups were diseased but no treatment-related pathology was identified. Based on the dose-related decrease in weight gain observed in females at 250 and 500 mg/kg diet, the NOEL in this study was 100 mg/kg diet, equal to 6 mg/kg bw per day. The author regarded the dose of 250 mg/kg diet as the non-toxic application rate. The reliability of any NOEL derived from this study should be considered questionable, owing to the presence of infection in all groups (Tettenborn & Hobik, 1968a; Trossmann & Hobik, 1970).

b) Dogs

Dogs of undetermined breed or source were given xylazine orally in gelatin capsules for 14-16 weeks at dose levels of 25, 50 or 100 mg/kg bw per day, 5 days/week. The low-dose group consisted of one male and one female, the mid-dose group of two males and the high-dose group of two males and two females. Haematology, clinical chemistry, blood coagulation, urinalysis and postmortem gross and microscopic evaluations were performed.

During week 8 of the test, one animal in the high-dose group died and was replaced with a new animal. Postmortem gross findings in this animal included diffuse reddening of the stomach and intestinal mucous membrane.

Histopathological findings in livers (fatty degeneration and necrosis) and kidneys (tubular epithelial necrosis and fat accumulation) of the high-dose group were considered treatment-related. Fatty deposits were noted in the liver and kidneys of the low-dose female. These findings were attributed to parturition, which occurred 3 weeks before the animal was killed. The author concluded that the NOEL for this study was 50 mg/kg bw per day. The reliability of any NOEL derived from this study should be considered questionable due to the lack of a control group and small numbers of animals in each test group (Tettenborn & Hobik, 1968b).

Beagle dogs (two/sex/group) were administered xylazine orally by dietary admixture for 13 weeks. Dosages administered were 0, 10, 30 or 100 mg/kg diet (equal to 0, 0.3, 0.9 or 3 mg/kg bw per day). At the beginning of the test the animals were approximately 8.5 months old and weighed 7-10 kg. Parameters evaluated included general appearance, ophthalmology, electrocardiography, haematology, clinical chemistry, urinalyses and gross and microscopic pathology.

No treatment-related adverse effects were observed in any of the parameters evaluated. The NOEL for this study was 3 mg/kg bw per day (Tettenborn, 1969; Mawdesley-Thomas, 1970).

2.2.2.2   2,6-Xylidine

a)   Rats

Three groups (nine or ten/group) of male Fischer-344 rats were given oral (gavage) doses of 160 mg 2,6-xylidine/kg bw per day for 5, 10 or 20 days. The dosage administered was 25% of the estimated LD_{50} determined by the investigator. A significant increase in splenic haemosiderosis (indicative of erythrocyte damage) after 20 days was noted as a treatment-related effect in this study. Splenic congestion and evidence for increased erythropoiesis were minimal (Short *et al.*, 1983).

Three groups of Sprague-Dawley rats (five/sex/group for controls, low and mid dose; four/sex/group for high dose) were administered 0, 20, 100 or 500-700 mg 2,6-xylidine/kg bw per day by gavage for 4 weeks. Treatment-related effects included decreased weight gain, decreased haemoglobin levels and hepatomegaly. In this study, the rat appeared to be about 10 times less susceptible to hepatotoxicity of 2,6-xylidine than the dog (see section 2.2.2.2b) (Magnusson *et al.*, 1971; IARC, 1993).

Two groups of 8-week-old Sprague-Dawley rats (5/sex/group) were orally administered (gastric intubation) a dose of 0 or 400 mg 2,6-xylidine per kg bw per day for 1 week immediately followed by a daily dose of 0 or 500 mg/kg bw for 3 weeks. Decreased body weight gain and hepatomegaly (most pronounced in centrilobular regions) were noted as treatment-related effects. Electron microscopy of liver tissue showed proliferation of hepatic smooth endoplasmic reticulum, which was deemed responsible for the observed hepatomegaly in treated rats. An increase in microsomal glucuronyltransferase was observed in males while aniline hydroxylase levels were increased in females. Decreases in liver glycogen and glucose-6-

phosphatase activity were also observed in the centrilobular regions of treated animals (Magnusson *et al.*, 1979).

Male Osborne-Mendel rats were administered up to 10 000 mg 2,6-xylidine per kg in the diet for 3-6 months. Treatment-related effects included 25% weight reduction, anaemia, hepatomegaly with no associated microscopic changes, splenic congestion and renal toxicity (Lindstrom *et al.*, 1963).

Groups of F-344/N rats (five/sex/group) were administered doses of 0, 80, 160, 310, 620 or 1250 mg/kg bw of 2,6-xylidine in corn oil by gavage 5 days/week for 2 weeks. Parameters evaluated included clinical observations, body weight, urinalysis, haematology, blood pH and carbon dioxide determinations, and gross postmortem findings.

Treatment-related deaths occurred at and above 620 mg/kg bw. All animals in the highest dose group died before the end of the study. A decrease of more than 10% in body weight was observed in males at and above 310 mg/kg bw and in females at and above 160 mg/kg bw. Generalized leukocytosis and an increase in the number of nucleated red blood cells were observed in male rats administered 310 or 620 mg/kg bw. Slight anisocytosis, poikilocytosis and polychromasia of the red blood cells occurred more frequently in dosed animals than in vehicle control animals. Moderate poikilocytosis occurred at 310 mg/kg bw and moderate polychromasia at 310 and 620 mg/kg bw. Slightly macrocytic erythrocytes were observed at the two highest doses. Slight anisocytosis, poikilocytosis and polychromasia were observed in female rats at 310 and 620 mg/kg bw. The NOEL for this study was 80 mg/kg bw per day (US National Toxicology Program, 1990).

Groups of F-344/N rats (10/sex/group) were given doses of 0, 20, 40, 80, 160 or 310 mg/kg bw of 2,6-xylidine in corn oil by gavage, 5 days/week for 13 weeks. Parameters evaluated included clinical observations, haematology, urinalysis, serum chemistry and enzyme analyses, gross and histopathological postmortem examinations.

A decrease in body weight gain of more than 10% occurred in males and females in the highest dose group and in females at 40 and 160 mg/kg bw per day. In the highest dose group, relative liver weights were significantly ($P = 0.003$) increased for males and females. Relative liver weight was also increased for males in the 160 mg/kg bw group. The liver weight to brain weight and kidney weight to brain weight ratios were significantly increased in females at 310 mg/kg bw per day.

Treatment-related effects on haematology included significantly decreased total leukocyte counts in males at doses of 40 mg/kg bw or more. These were accompanied by decreases in the percentage of lymphocytes and increases in the percentage of segmented neutrophils at doses of 80 mg/kg bw or more. In males, haemoglobin levels were significantly decreased at

160 and 310 mg/kg bw and erythrocyte and haematocrit levels were decreased at 310 mg/kg bw. The NOEL for this study was 20 mg/kg bw per day (US National Toxicology Program, 1990).

b)   Dogs

Four groups of beagle dogs (one/sex/group) were given an oral (gelatin capsule) dose of 0, 2, 10 or 50 mg 2,6-xylidine/kg bw per day for 4 weeks. Treatment-related effects included vomiting (mid- and high-dose groups), poor condition and decreased body weights (high-dose group), hyperbilirubinaemia (mid- and high-dose groups), hypoproteinaemia (mid- and high-dose groups) and fatty degenerative changes in the liver that increased in severity with increasing dose (Magnusson et al., 1971; IARC, 1993).

2.2.3   Long-term toxicity/carcinogenicity studies

2.2.3.1   Xylazine

No carcinogenicity studies have been performed with xylazine

2.2.3.2   2,6-Xylidine

Four groups of Charles River CRL:COBS CD (SD) BR rats (56/sex/group) were fed diets containing 2,6-xylidine (99.06% pure) at concentrations of 0, 300, 1000 or 3000 mg/kg diet (equivalent to 0, 15, 50 or 150 mg/kg bw per day) for 102 weeks. The animals assigned to this study were F_{1a} generation weanlings from a multigeneration study in which animals were fed diets containing 0, 300, 1000 or 3000 mg/kg 2,6-xylidine beginning at 5 weeks of age. Parameters evaluated in the carcinogenicity study included clinical observations, haematology, blood urea nitrogen, glucose, SGOT, alkaline phosphatase and gross and microscopic postmortem examinations.

Treatment-related clinical effects included a decrease in mean body weight gain in high-dose males and females ($> 10\%$). Mortality was significantly ($P < 0.001$) increased (relative to controls) in males in the high-dose group. Mortality was also increased for mid-dose males. Survival at 105 weeks was 43/56, 40/56, 33/56 and 14/56 for males in the control, low-, mid- and high-dose groups, respectively. For females, survival was 33/56, 25/56, 32/56 and 24/56 for the controls, low-, mid- and high-dose groups, respectively.

Microscopically, a significant increase in carcinoma of the nasal cavity was observed in high-dose males (26/56; $P < 0.001$, life table test). For females, the incidence of carcinomas of the nasal cavity were 0/56, 0/56, 1/56 and 24/56 in the low-, mid- and high-dose groups, respectively ($P < 0.001$, life table test). Two adenocarcinomas were diagnosed in high-dose males. The incidence of papillary adenomas in males was 0/56 in controls, 0/56 in low-dose, 2/56 in mid-dose and 10/56 in high-dose rats (P = 0.001, incidental tumour test). For females, nasal adenomas occurred in 0/56 in controls, 0/56 in low-dose, 1/56 in mid-dose and 6/56 in high-dose rats (P = 0.02, incidental tumour test). Several unusual neoplasms of the nasal cavity were also considered to be related to treatment. These included one undifferentiated sarcoma identified in one high-dose female, rhabdomyosarcomas which occurred in two high-dose male and two high-dose females and malignant mixed tumours having features associated with both adenocarcinoma and rhabdomyosarcoma were observed in one high-dose male and one high-dose female rat. Non-neoplastic nasal cavity lesions included acute inflammation (rhinitis), epithelial hyperplasia and squamous metaplasia. These occurred at increased incidence (relative to controls) in high-dose male and female rats. The incidence of subcutaneous fibromas and fibrosarcomas combined in males was 0/56, 2/56, 2/56 and 5/56 for the control, low-, mid- and high-dose groups, respectively (P = 0.001, life table test; P < 0.001 life table trend test). For females, the incidence of these tumours combined was 1/56, 2/56, 2/56 and 6/56 for controls, low-, mid- and high-dose groups, respectively ((P = 0.01, life table trend test). Neoplastic nodules occurred in livers of female rats with a significant positive trend. The incidence was 0/56, 1/56, 2/56 and 4/55 for the controls, low-, mid-, and high-dose groups, respectively (P = 0.03, incidental test; P = 0.012, incidental trend test).

Treatment-related effects on haematology included decreases in erythrocyte counts and haemoglobin levels at 18 months in the high-dose males. Decreases in these parameters were also observed in the mid- and high-dose females at 12 months. The author remarked that these changes were not severe enough to be considered indicative of anaemia.

The author concluded that under the conditions of this study, 2,6-xylidine was clearly carcinogenic for male and female Charles River CD rats. This was based on the observed significant increases in the incidence of adenomas and carcinomas of the nasal cavity. Additionally the author stated that the increased incidence of subcutaneous fibromas and fibrosarcomas in male and female rats and increased incidence of neoplastic nodules of the liver in female rats could have been treatment-related (US National Toxicology Program, 1990).

The International Agency for Research on Cancer (IARC) has evaluated the carcinogenic risk of 2,6-xylidine to humans. The Working Group concluded that there was inadequate evidence in humans but sufficient evidence in experimental animals for the carcinogenicity of 2,6-xylidine. The IARC classified 2,6-xylidine as Group 2B (possibly carcinogenic to humans) (IARC, 1993).

2.2.4 Special studies on teratogenicity

Xylazine was administered by gavage to groups of pregnant rats (22 animals/group) on gestation days 6 to 15, then killed on day 20 for examination of uterine contents. Dosages administered were 0, 1, 4 or 16 mg/kg bw per day. The study was conducted in accordance with the principles of Good Laboratory Practice Standards and Guidelines of the OECD, United Kingdom, FDA and Japan.

Treatment-related maternal effects included partial closing of the eyelids, underactivity, ataxia, flat posture and slightly reduced body weight gain in the high-dose group only. Fetal effects included a decrease in mean fetal weight in the high-dose group only. A teratogenic potential of xylazine was not evident at levels up to and including 16 mg/kg bw per day. The NOEL in this study was 4 mg/kg bw per day (Reynolds, 1994).

2.2.5 Special studies on genotoxicity

The results of genotoxicity studies with xylazine and 2,6-xylidine are summarized in Table 4.

The results of bacterial mutagenicity testing of xylazine were considered to be negative by the author. Reviewing the data, the Committee concluded that a more than two-fold reproducible increase in revertant colonies in tester strains TA1535 and TA1538 represents weak mutagenic activity, even in the absence of a clear dose-response.

2.2.6 Special studies on methaemoglobin and haemoglobin adduct formation with 2,6-xylidine

2.2.6.1 Cats and dogs

Cats and dogs (numbers not specified) were administered an i.v. dose of 30 mg 2,6-xylidine/kg bw or an oral dose of 164 mg *N*-acetyl 2,6-xylidine/kg bw. 2,6-Xylidine induced a 10% methaemoglobinaemia in cats, and *N*-acetyl 2,6-xylidine induced a 5% methaemoglobinaemia in cats. Haemoglobin was unaffected in dogs in this study (McLean *et al.*, 1967).

Five adult cats (>24 months old) were administered an i.v. dose of 30 mg 2,6-xylidine/kg bw. Blood samples were drawn at 1, 2, 3, 4 and 5 hours after dosing and analysed for methaemoglobin formation. The mean methaemoglobin concentration determined from these sampling intervals was 7% (range = 4.8% to 8.7%). Prior to treatment the mean methaemoglobin concentration of the 152 cats used in this study was approximately 1% (Mclean *et al.*, 1969).

Table 4. Genotoxicity assays with xylazine and 2,6-xylidine

Test system	Test object	Concentration	Results	Reference
Xylazine				
In vitro				
Reverse mutation[1]	*S. typhimurium* TA1535,	0.4-12 mg/plate	Weak positive (-S9)	Herbold, 1984
	TA1538	0.4-12 mg/plate	Weak positive (-S9)	Herbold, 1984
	TA100,	0.4-12 mg/plate	Negative	Herbold, 1984
	TA98,	0.4-12 mg/plate	Negative	Herbold, 1984
	TA1537	0.4-12 mg/plate	Negative	Herbold, 1984
Mammalian cell forward mutation[1]	V79/HGPRT	62-1250 µg/ml (-S9) 2-40 µg/ml (+S9)	Negative	Brendler-Schwaab, 1994
In vivo				
Cytogenetic assay	Mouse bone marrow	50 mg/kg bw, i.p.[2]	Negative	Herbold, 1995
2,6-Xylidine				
In vitro				
Reverse mutation[1]	*S. typhimurium* TA1535	100-9900 µg/plate	Negative	US National Toxicology Program, 1990
		3 µmol/plate	Negative	Florin *et al.*, 1980[4]
		0.1-10 mg/plate	Negative	Zeiger *et al.*, 1988
	TA100	100-9900 µg/plate	Negative	US National Toxicology Program, 1990
		360 µg/plate	Negative	Florin *et al.*, 1980[4]
		0.1-10 mg/plate	Neg (-S9), Pos (+S9)	Zeiger *et al.*, 1988[3]
		480-4000 µg/plate	Negative	Kugler-Steigmeier *et al.*, 1989
	TA1537	100-9900 µg/plate	Negative	US National Toxicology Program, 1990
		360 µg/plate	Negative	Florin *et al.*, 1980[4]
		0.1-10 mg/plate	Negative	Zeiger *et al.*, 1988

Table 4 (continued).

Test system	Test object	Concentration	Results	Reference
	TA98	100-9900 μg/plate	Negative	US National Toxicology Program, 1990
		360 μg/plate	Negative	Florin et al., 1980[4]
		0.1-10 mg/plate	Negative	Zeiger et al., 1988
Gene mutation[1]	Mouse lymphoma L5178Y cells, tk locus	Not given	Positive	Rudd et al., 1983[7]
Sister chromatid exchange[1]	Chinese hamster ovary cells	30-1500 μg/ml	Positive	Galloway et al., 1987
In vivo Cytogenetic assay	ICR mouse bone marrow	350 mg/kg bw, p.o.	Inconclusive[6]	Parton et al., 1988
		375 mg/kg bw, p.o.	Inconclusive[6]	Parton et al., 1990
In vivo-in vitro DNA repair assay	Rat primary hepatocytes	40-850 mg/kg bw, p.o.	Negative	Mirsalis et al., 1989
Covalent DNA binding	Rats	87.2 μCi ^{14}C-labelled 2,6-xylidine/rat, i.p.[5]	Positive	Short et al., 1989

[1] Both with and without rat liver S9 fraction
[2] Cyclophosphamide positive control
[3] Weakly positive in two of three laboratories, negative in the third
[4] Spot tests only
[5] Pretreatment with unlabelled 262.5 mg/kg bw 2,6-xylidine daily for 9 days
[6] Results suggest test article may not have reached target tissue (bone marrow)
[7] Reference was an abstract and doses were not stated in that reference

2.2.6.2   Humans

2,6-Xylidine-haemoglobin adduct levels have been found to be elevated in human patients receiving lidocaine treatment for local anaesthesia (1 mg/kg bw) or cardiac arrhythmias (up to 50 mg/kg bw, i.v.). 2,6-Xylidine-haemoglobin adducts are also found in humans with no known exposure to lidocaine. This is attributed to the 120-day lifespan of the erythrocyte and chronic exposure to environmental or iatrogenic sources of aromatic amines, e.g., cigarette smoke. The levels of 2,6-xylidine-haemoglobin adducts found correspond to at an estimated daily exposure (from iatrogenic and environmental sources) of 23 µg (IARC, 1993; Bryant et al., 1994).

Methaemoglobinaemia induced by i.v. administration of lidocaine was studied in 40 human cardiac patients. Treatment consisted of a 1 mg/kg bw i.v. bolus followed 15 minutes later with a 0.5 mg/kg bw i.v. bolus. Patients were maintained between and after bolus doses with an infusion rate of 1-4 mg lidocaine/min. Blood samples were drawn before treatment and 1 and 6 hours after treatment. Although the investigators found methaemoglobin levels in these patients to be significantly elevated, the increase was not large enough to be of clinical concern. The highest methaemoglobin level attained was 1.2%. The author did not address the possible role of the 2,6-xylidine metabolite in the observed increases in methaemoglobin levels in treated patients (Weiss et al., 1987).

2.3   Observations in humans

A 34-year-old man self-injected 10 ml of a 100 mg/ml solution of xylazine intramuscularly. The estimated dose was 15 mg/kg bw. The individual was discovered (30 minutes after retiring for bed) in a deeply comatose, apnoeic and areflexic state. An empty Rompun (xylazine) bottle (known to have contained 10 cc earlier) lay by his side. He was immediately admitted to the hospital. Upon admission his pupils were of moderate size and responded slowly to light. Other findings included a blood pressure of 120/70 mmHg and heart rate of 60 bpm with stable sinus rhythm. Lactate dehydrogenase (LDH) activity was elevated, with the LDH-1 isoenzyme predominating. Creatine phosphokinase (CPK) activity was also elevated, particularly in the CPK-3 and CPK-2 isoenzymes. These enzyme changes persisted for 5 to 7 days. Plasma glucose level was also elevated. Two days following hospital admission sinus tachycardia developed, interspersed with runs of multifocal premature ventricular contractions which were controlled with lidocaine infusion. Blood pressure remained approximately 120/80 for the duration of hospitalization. Coma and respiratory depression lasted 60 hours. The patient was discharged from the hospital 17 days after admission.

The author noted that the patient might have died if he had not been found shortly after the injection was administered, owing to the marked respiratory depression that occurred. Hypotension, which has been reported as an effect of xylazine in humans, did not occur in this case. According to the author, the enzyme activity increases indicated that myocardial muscle damage had occurred and an intrinsic cardiotoxic effect of xylazine was suspected. Finally, in this case the greatest threat to life was the CNS-depressant effect of xylazine (Carruthers et al., 1979).

A 20-year-old woman ingested 400 mg of xylazine. Approximately 2 hours later she became drowsy, incontinent (urine), difficult to arouse and occasionally unresponsive to verbal commands. She was admitted to the hospital approximately 3 hours after the ingestion. In a similar way to the case of human poisoning described by Carruthers *et al.* (1979), she experienced a relatively low initial cardiac rate, which later gradually increased, significant central nervous system and respiratory depression, transient hyperglycaemia and ventricular arrhythmias. However there was no evidence of myocardial damage. A sample of this patient's urine was analysed using a gas chromatograph-mass spectrometer computer system. Xylazine was found largely unchanged in the urine as shown by a lack of any structurally related compounds in the basic urine extract. No xylazine was found in a blood sample taken at the time of admission. The author concluded that plasma levels were below the limit of detection of the method (100 ng/ml). The patient was discharged ambulatory and without apparent adverse effects two days following admission to the hospital (Gallanosa *et al.*, 1981).

A 36-year-old man died following ingestion of alcohol and clorazepate combined with an injection of approximately 40 ml of xylazine (100 mg/ml). Xylazine was found in the decedent's blood, brain, kidney, liver, lung, fat and urine at concentrations of 0.2, 0.4, 0.6, 0.9, 1.1, 0.05 and 7 ppm, respectively (Poklis *et al.*, 1985).

A 29-year-old woman self-injected 40 mg of xylazine intramuscularly. The estimated dose was 0.73 mg/kg bw. Clinical findings included disorientation, miosis, hypotension and bradycardia, but no cardiac arrhythmias were noted. The abnormalities resolved spontaneously (Spoerke *et al.*, 1986).

A 37-year-old woman self-injected 24 ml (2400 mg) xylazine intramuscularly. The estimated dose was 22 mg/kg bw. Twenty minutes after the injection her blood pressure was 166/130 mmHg, heart rate was 76 bpm and respirations 18 per minute. The serum glucose level was 175 mg/dl. Blood pressure later decreased to 130/90 mmHg and she became apnoeic. No cardiac arrhythmias were observed during her 3 days of hospitalization. Hypotension and bradycardia occurred two days after the injection. The patient survived (Spoerke *et al.*, 1986).

A 29-year-old woman self-injected an unknown amount of xylazine intravenously. She became apnoeic and had an initial blood pressure of 130/90 mmHg with a pulse of 60 bpm. Serum glucose levels did not exceed 90 mg/dl. Twenty-four hours after the injection the patient experienced hypotension and bradycardia. Spontaneous respiration resumed 18 hours after hospital admission. The patient recovered fully (Spoerke *et al.*, 1986).

A 19-year-old man accidentally injected himself subcutaneously with 2 ml (100 mg/ml) of xylazine. The dose administered was 3 mg/kg bw. Thirty minutes later he became difficult to rouse and was hospitalized. Clinical findings included miosis, hyporeflexia, hypotension, bradycardia, respiratory and central nervous system depression and hyperglycaemia. He was treated with intravenous fluids and assisted ventilation. Eight hours after hospitalization the patient was alert and responsive. Twenty-four hours later he was released (Samanta *et al.*, 1990).

A 39-year-old woman was admitted to the hospital with symptoms of tiredness, faintness and blurred vision. Clinical findings included sinus bradycardia with a blood pressure of 130/90. Xylazine was found in the urine and serum at concentrations of 1674 μg/litre and 30 μg/litre, respectively (Lewis *et al.*, 1983).

3.   COMMENTS

The Committee considered toxicological data on xylazine, including the results of acute and short-term toxicity studies as well as studies on pharmacodynamics, pharmacokinetics, reproductive and developmental toxicity, genotoxicity and effects in humans. In addition, toxicological studies on 2,6-xylidine, a metabolite of xylazine, were reviewed; these included studies on acute and short-term toxicity, carcinogenicity and genotoxicity.

Numerous pharmacological side-effects of xylazine have been observed in treated animals, including mydriasis, impairment of thermo-regulatory control, various effects on the cardiovascular system, acid-base balance and respiration, hyperglycaemia, and haematological and gastro-intestinal effects. Cattle and sheep are approximately 10 times more sensitive to xylazine than horses, dogs and cats.

Rats were administered radiolabelled xylazine intravenously at doses of 0.02 to 10 mg/kg bw or orally at doses of 0.02 to 100 mg/kg bw. More than 95% of the oral dose was absorbed, with a half-life of approximately 5 minutes. Following oral or intravenous administration, approximately 70% of the administered dose was eliminated in urine and 30% in faeces. Renal excretion following oral or intravenous administration was associated with a half-life of 2 to 3 hours. Enterohepatic circulation did not occur to a notable extent. In cattle administered an intramuscular dose of 0.2 or 0.5 mg xylazine/kg bw, less than 1% of the dose was excreted unchanged in the urine, and the parent compound was detected in the urine up to 6 hours following administration. Metabolites of xylazine were detected in urine from these cattle up to 10 hours following administration.

Pharmacokinetic parameters following intravenous administration showed minor variations between species. Xylazine disappeared rapidly from plasma following intravenous administration, with an elimination half-life of approximately 40 minutes in cattle and approximately 20 minutes in sheep. Xylazine could not be detected in the plasma of cattle following intramuscular administration of a single therapeutic dose.

In rats administered an intravenous dose of 2 mg/kg bw radio-labelled xylazine, approximately 20 metabolites were quantified as xylazine equivalents in urine and bile. The major metabolite comprised 35% of the administered dose. Approximately 8% of the dose was eliminated as unchanged xylazine 24 hours after dosing. In an *in vitro* study, 4 metabolites were identified when xylazine was incubated with rat liver microsomes. The same metabolites were identified in the urine of horses treated with xylazine. The major metabolite in both cases was identified as N-(2,6-dimethylphenyl)thiourea. In cattle administered an intramuscular dose of 0.2 mg xylazine/kg bw (two cows) or 0.5 mg xylazine/kg bw (one cow), 2,6-xylidine was identified as a metabolite excreted in urine in both conjugated and unconjugated forms.

The acute oral toxicities of xylazine and 2,6-xylidine were tested in mice and rats. Xylazine was determined to be moderately toxic (LD_{50} = 121-240 mg/kg bw) and 2,6-xylidine to be slightly toxic (LD_{50} = 600-1000 mg/kg bw).

Three studies on the short-term toxicity of xylazine were reviewed. A 32-week dietary study in rats and a 16-week oral (capsules) study in dogs were considered inadequate for the determination of the toxicity of xylazine owing to the use of insufficient numbers, poor quality animals and inadequate study design. The third was a 13-week oral study in beagle dogs fed diets containing 0, 10, 30 or 100 mg/kg xylazine in the feed (equal to 0.3, 0.9 or 3 mg/kg bw per day). No treatment-related effects were observed in any of the treated groups.

In a two-week oral (gavage) toxicity study in rats with 2,6-xylidine, rats were dosed with 80, 160, 310, 620 or 1250 mg 2,6-xylidine/kg bw per day, 5 days per week. Treatment-related effects included increased mortality (all animals in the high-dose group died), decreased body weight (males at 310 mg/kg bw per day and above and females at 160 mg/kg bw per day and above) and various effects on haematological parameters as indicated by leukocytosis and changes in red blood cell parameters indicative of increased erythropoiesis (males and females at 310 mg/kg bw per day and above). The NOEL in this study was 80 mg/kg bw per day.

In a 13-week oral (gavage) toxicity study in rats with 2,6-xylidine, rats were dosed with 20, 40, 80, 160 or 310 mg 2,6-xylidine/kg bw per day,

5 days per week for 13 weeks. Treatment-related effects included decreased body weight gain (males at 310 mg/kg bw per day and females at 40 mg/kg bw per day and above), increased absolute and relative liver weights (females at 160 mg/kg bw per day and above; males at 310 mg/kg bw per day), leukopenia (males at 40 mg/kg bw per day and above), haemoglobinaemia (males at 160 mg/kg bw per day and above) and anaemia (males at 310 mg/kg bw per day). The NOEL was 20 mg/kg bw per day.

In a carcinogenicity study, male and female rats were fed diets containing 2,6-xylidine at concentrations of 300, 1000 or 3000 mg/kg food (equivalent to 15, 50 or 150 mg/kg bw per day). Significant increases in the incidences of papillomas and carcinomas of the nasal cavity were observed in high-dose males and females. There was a significant dose-related increase in the incidence of adenomas in the nasal cavity in both males and females. In addition, unusual rhabdomyosarcomas and malignant mixed tumours of the nasal cavity were observed in the high-dose males and females. There was a dose-related significant increase in the incidence of subcutaneous fibromas and fibrosarcomas in both treated males and females. In females, neoplastic nodules occurred in livers with a significant positive trend and the increase was significant in the high-dose group by the incidental tumour test. The Committee concluded that 2,6-xylidine was carcinogenic in this study.

The International Agency for Research on Cancer has evaluated the carcinogenic risk of 2,6-xylidine and has classified it as Group 2B (possibly carcinogenic to humans).

In a teratogenicity study, xylazine was administered to pregnant rats at doses of 1, 4 or 16 mg xylazine/kg bw per day on gestation days 6 to 15. Treatment-related maternal effects included partial closing of the eyelids, hypoactivity, ataxia, flat posture and slightly reduced body weight gain in the high-dose group only. A decrease in mean fetal weight was seen in the high-dose group. No teratogenic effects were noted in this study. The NOEL for maternal and fetal effects was 4 mg/kg bw per day.

Xylazine has been tested in reverse mutation assays in *Salmonella*, a forward mutation assay in cultured mammalian cells and in an *in vivo* cytogenetic assay. In *Salmonella*, weak positive results were obtained. Negative results were observed in a forward mutation assay on cultured mammalian cells and in a mouse bone marrow micronucleus test. The Committee concluded that xylazine is weakly mutagenic.

2,6-Xylidine was tested in a series of *in vitro* and *in vivo* genotoxic assays. It was weakly positive for reverse mutation in *Salmonella*. In mammalian cells, it induced forward mutation and was positive in a sister chromatid exchange test. Inconclusive results were obtained in a mouse bone marrow micronucleus test because there was no assurance that the bone marrow had been adequately exposed. 2,6-Xylidine was found to be inactive in an *in vivo-in vitro* rat hepatocyte unscheduled DNA synthesis assay. Covalent binding of the compound to DNA was observed in rats. The Committee concluded that 2,6-xylidine is genotoxic.

The potential for 2,6-xylidine to induce methaemoglobinaemia was reviewed by the Committee. Single doses of 30 mg 2,6-xylidine/kg bw intravenously or 164 mg/kg bw *N*-acetyl-2,6-xylidine orally have been shown to induce methaemoglobinaemia in cats but not in dogs. 2,6-Xylidine has also been shown to be a product of lidocaine metabolism in humans. Methaemoglobin and 2,6-xylidine-haemoglobin adduct levels have been shown to increase in human cardiac patients receiving lidocaine treatment.

Effects of xylazine on humans poisoned following accidental or intentional self-injection (0.7-15 mg/kg bw) or ingestion (7 mg/kg bw) included symptoms of central nervous system depression, respiratory depression, hypo- and hypertension, bradycardia, tachycardia, ventricular arrhythmias, and transient hyperglycaemia.

4.   EVALUATION

The Committee was unable to establish an ADI for xylazine because it concluded that the 2,6-xylidine metabolite was genotoxic and carcinogenic. Annex 4 lists the information that would be required for further review.

5.   ACKNOWLEDGMENTS

The preparer of the first draft would like to recognize the following individuals for their assistance and contributions to the preparation of the first draft:

> Ms. Deborah Brooks, information specialist, Center for Veterinary Medicine
> Dr. Steve Brynes, residue chemist, Center for Veterinary Medicine
> Dr. Jennifer Burris, veterinary pathologist, Center for Veterinary Medicine
> Dr. Robert Condon, biostatistician, Center for Veterinary Medicine
> Dr. Haydee Fernandez, toxicologist, Center for Veterinary Medicine
> Dr. Devaraya Jagannath, genetic toxicologist, Center for Veterinary Medicine
> Dr. Alan Pinter, toxicologist, National Institute of Public Health, Budapest, Hungary
> Dr. Leonard Schechtman, genetic toxicologist, Center for Veterinary Medicine

6. REFERENCES

Amer, A.A. & Misk, N.A. (1980). Rompun in goats with special reference to its effect on the cerebral spinal fluid (c.s.f.). *Vet. Med. Rev.*, **2**, 168-174.

Bauman, E.K. & Nelson, D.L. (1969). Toxicity of BAY Va 1470 to cats. Unpublished report No. 24208 from Chemagro Corporation. Submitted to WHO by Bayer AG, Leverkusen, Germany.

Benson, G.J., Thurmon, J.C., Neff-Davis, C.A., Corbin, J.E., Davis, L.E., Wilkinson, B., & Tranquilli, W.J. (1984). Effect of xylazine hydrochloride upon plasma glucose concentrations in adult pointer dogs. *J. Am. Anim. Hosp. Assoc.*, **20**, 791-794.

Bolte, S. & Stupariu, A (1978). Motility of the rumen in cattle and sheep under the influence of neuroleptics and analgesics, particularly xylazine, propionylpromazine and diazepam. *Lucrari Stiintifice Inatitutul Agron. Timisoara, Ser. Med. Vet.*, **15**, 157-167.

Booth, N.H. (1988). Nonnarcotic analgesics. In: Booth N.H.& McDonald E. (eds.), Veterinary Pharmacology and Therapeutics, 6th edition, Iowa State University Press, Ames, Iowa, pp. 351-359.

Brendler-Schwaab, S. (1994). Rompun hydrochloride. Mutagenicity study for the detection of induced forward mutations in the V79-HGPRT assay *in vitro*. Unpublished report No. T9049349 from Bayer AG, Fachbereich Toxicology. Submitted to WHO by Bayer AG, Leverkusen, Germany.

Bryant, M.S., Simmons, H.F., Harrell, R.E., & Hinson, J.A. (1994). 2,6-Dimethylaniline-hemoglobin adducts from lidocaine in humans. *Carcinogenesis*, **15**(10), 2287-2290.

Carruthers, S.G., Nelson, M., Wexler H.R., & Stiller, C.R. (1979). Xylazine hydrochloride (Rompun) iverdise in man. *Clin. Toxicol.*, **15**(3), 281-285.

Carter, S.W., Robertson, S.A., Steel, C.J., & Jourdenais, D.A. (1990). Cardiopulmonary effects of xylazine sedation in the foal. *Equine Vet. J.*, **22**(6), 384-388.

Colby, E.D., McCarthy, L.E., & Borison, H.L. (1981). Emetic action of xylazine on the chemoreceptor trigger zone for vomiting in cats. *J. Vet. Pharmacol. Ther.*, **4**, 93-96.

Crawford, C.R., Nelson, D.L., & Anderson, R.H (1970a). The toxicity of BAY Va 1470 2% injectable to dogs and cats given at ten times recommended dose. Unpublished report No. 28139 from Research

Department, Chemagro Corporation. Submitted to WHO by Bayer AG, Leverkusen, Germany.

Crawford, C.R., Nelson, D.L., & Anderson, R.H (1970b). The toxicity of BAY Va 1470 10% injectable alone and in combination with ®NEGUVON* soluble powder (batch No. 0050337) to horses. Unpublished report No. 28136 from Research Department, Chemagro Corporation. Submitted to WHO by Bayer AG, Leverkusen, Germany.

Cullen, L.K. & Jones, R.S. (1977). Clinical observations on xylazine/ketamine anaesthesia in the cat. *Vet. Rec.*, **101**, 115-116.

Demoor, A. & Desmet, P. (1971). Effect of Rompun on acid-base equilibrium and arterial oxygen pressure in cattle. *Vet. Med. Rev.*, **2-3**, 163-169.

Duhm, B., Maul, W., Medenwald, H., Patzschke, K., & Wegner, L.A. (1968). Experimental tests on animals with radioactive labled BAY Va 1470. Unpublished report No. 22874 from Bayer Isotope Institute, Elberfeld Branch. Submitted to WHO by Bayer AG, Leverkusen, Germany.

Duhm, B., Maul, W., Medenwald, H., Patzschke, K., & Wegner, L.A. (1969). Experiments using radioactively tagged BAY Va 1470 on rats. *Berl. Münch. Tierärztl. Wochenschr.*, **82**, 104-109.

Eichner, R.D., Prior, R.L., & Kvasnicka, W.G (1979). Xylazine-induced hyperglycemia in beef cattle. *Am. J. Vet. Res.*, **40**(1), 127-129.

Feldberg, W. & Symonds, H.W. (1980). Hyperglycemic effect of xylazine. *J. Vet. Pharmacol. Ther.*, **3**, 197-202.

Florin, I., Rutberg, L., Curvall, M., & Enzell, C.R. (1980). Screening of tobacco smoke constituents for mutagenicity using the Ames' test. *Toxicology*, **15**, 219-232.

Freire, A.C.T, Gontijo, R.M., Pessoa, J.M., & Souza, R. (1981). Effect of xylazine on the electrocardiogram of sheep. *Br. Vet. J.*, **137**, 590-595.

Gallanosa, A.G., Spyker, D.A., Shipe, J.R., & Morris, D.L. (1981). Human xylazine overdose: A comparative review with clonidine, phenothiazines, and tricyclic antidepressants. *Clin. Toxicol.*, **18**(6), 663-678.

Galloway, S.M., Armstrong, M.J., Reuben, C., Colman, S., Brown, B., Ccannon C., Bloom, A.D., Nakamura, F., Ahmed, N., Duk, S., Rimpo, J., Margolin, B.H., Resnick, M.A., Anderson, B., & Zeiger, E. (1987). Chromosome aberrations and sister chromatid exchanges in Chinese hamster ovary cells: evaluations of 108 chemicals. *Environ. Mol. Mutagen.*, **10**(10), 1-175

Garcia-Villar, R., Toutain, P.L., Alvinerie, M., & Ruckenbusch Y. (1981). The pharmacokinetics of xylazine hydrochloride: an interspecific study. *J. Vet. Pharmacol. Ther.*, **4**, 87-92.

Gross, M.E. & Tranquilli, W.J. (1989). Use of alpha-2-adrenergic receptor antagonists. *J. An. Vet. Med. Assoc.*, **195**(3), 378-381.

Herbold, B. (1984). Salmonella/microsome test to evaluate for point-mutagenic effects. Unpublished report No. T0016805 from Bayer Sparte Pharma. Submitted to WHO by Bayer AG, Leverkusen, Germany.

Herbold, B. (1995). Rompun Hydrochloride. Micronucleus test on the mouse. Unpublished report No. T9058169 + T2058171 from Bayer AG Fachbereich Toxicology. Submitted to WHO by Bayer AG, Leverkusen, Germany.

Hikasa, Y. Takase, K., & Ogasawara, S. (1989). Evidence for the involvement of alpha-2-adrenoceptors in the emetic action of xylazine in cats. *Am. J. Vet. Res.*, **50**, 1348-1351.

Hikasa, Y., Takase, K., Osada, T., Takamatsu, H., & Ogasawara, S. (1987). Xylazine-induced vomiting in dogs: elimination by ablation of the area postrema and blockade by yohimbine. *Zent.bl. Vet.med.*, **A34**(2), 154-158.

Holmes, A.M. & Clark, W.T. (1977). Xylazine sedation of horses. *N.Z. Vet. J.*, **25**, 159-161.

Hsu, W.H. & Hummel, S.K. (1981). Xylazine-induced hyperglycemia in cattle: a possible involvement of alpha-2-adrenergic receptors regulating insulin release. *Endocrinology*, **109**, 825-829.

Hsu, W.H. & McNeel, S.V. (1983). Effect of yohimbine on xylazine-induced prolongation of gastrointestinal transit in dogs. *J. Am. Vet. Med. Assoc.*, **183**(3), 297-300.

Hsu, W.H., Betts, D.H., & Lee, P. (1981). Xylazine-induced mydriasis: possible involvement of a central postsynaptic regulation of parasympathetic tone. *J. Vet. Pharmacol. Ther.*, **4**, 209-214.

Hsu, W.H., Hanson, C.E., Hembrough, F.B., & Schaffer, D.D. (1989). Effects of idazoxan, tolazoline and yohimbine on xylazine induced respiratory changes and central nervous system depression in ewes. *Am. J. Vet. Res.*, **50**(9), 1570-1573.

IARC (1993). 2,6-Dimethylaniline (2,6-xylidine). IARC Monograph on the Evaluation of Carcinogenic Risks to Humans: Occupational exposures of hairdressers and barbers and personal use of hair colourants; some hair dyes, cosmetic colourants, industrial dyestuffs and aromatic amines, **57**, 323-335.

Jacobson, K.H. (1972). Short communication. Acute oral toxicity of mono- and di-alkyl ring-substituted derivatives of aniline. *Toxicol. Appl. Pharmacol.*, **22**, 153-154.

Klide, A.M., Calderwood, H.W., & Simen L.R. (1975). Cardio-pulmonary effects of xylazine in dogs. *Am. J. Vet. Res.*, **36**(7), 931-935.

Kugler-Steigmeier, M.E., Frierich, U., Graf, U., Lutz, W.K., Maier, P., & Schlatter, C. (1989). Genotoxicity of aniline derivatives in various short-term tests. *Mutat. Res.*, **211**, 279-289.

Leblanc, P.H. & Eberhart, S.W. (1990). Cardiopulmonary effects of epidurally administered xylazine in the horse. *Equine Vet. J.*, **22**(6), 389-391.

Lewis, S., O'Callaghan, C.L.P., & Toghill, P.J. (1983). Clinical curio: self medication with xylazine. *Br. Med. J.*, **287**, 1369.

Lindstrom, H.V., Hansen, W.H., Nelson, A.A., & Fitzhugh, O.G. (1963). The metabolism of FD&C Red No. 1. II. The fate of 2,5-*para*-xylidine and 2,6-*meta*-xylidine in rats and observations of the toxicity of xylidine isomers. *J. Pharmacol. Exp. Ther.*, **142**, 257-264.

Lindstrom, H.V., Bowie, W.C., Wallace, W.C., Nelson, A.A., & Fitzhugh, O.G. (1969). The toxicity and metabolism of mesidine and pseudocumidine in rats. *J. Pharmacol. Exp. Ther.*, **167**(2), 223-234.

McLean, S., Murphy, B.P., Starmer, G.A., & Thomas, J. (1967). Methaemoglobin formation induced by aromatic amines and amides. *J. Pharm. Phamacol.*, **19**, 146-154.

McLean, S., Starmer, G.A., & Thomas, J. (1969). Methaemoglobin formation by aromatic amines. *J. Pharm. Pharmacol.*, **21**, 441-450.

Magnusson, G., Bodin, N.-O., & Hansson, E. (1971). Hepatic changes in dogs and rats induced by xylidine isomers. *Acta Pathol. Microbiol. Scand.*, **79**, 639-648.

Magnusson, G., Majeed, S.K., Down, W.H., Sacharin, R.M., & Jorgeson, W. (1979). Hepatic effects of xylidine isomers in rats. *Toxicology*, **12**, 63-74.

Mawdesley-Thomas, L.E. (1970). Pathology report of the subchronic toxicity for dogs by oral administration of compound BAY Va 1470. Unpublished report No. 26692 from Huntingdon Research Centre, Huntingdon, England. Submitted to WHO by Bayer AG, Leverkusen, Germany.

Mirsalis, J.C., Tyson, C.K., Steinmetz, K.L., Loh, E.K., Hamilton, C.M., Bakke, J.P., & Splading, J.W. (1989). Measurement of unscheduled DNA synthesis and S-phase synthesis in rodent hepatocytes following *in vivo* treatment: testing of 24 compounds. *Environ. Mol. Mutagen.*, **13**, 155-164.

Murphy, J.J. & Jacobs, K. (1975). Residues of ROMPUN and its metabolites in cattle. Unpublished report No. 43814 from Chemagro Agricultural Division, Mobay Chemical Corporation. Submitted to WHO by Bayer AG, Leverkusen, Germany.

Mutlib, A.E., Chui, Y.C., Young, L.M., & Abbott, F.S. (1992). Characterization of metabolites of xylazine produced *in vivo* and *in vitro* by LC/MS/MS and by GC/MS. *Drug Metab. Dispos.*, **20**(6), 840-848.

Nelson, D.L., Bauman, E.K., Mosier, J.O., & Allen, A.D. (1968a). A study of the effect of large doses of BAY Va 1470 to horses. Unpublished report No. 23405 from Research and Development Department, Chemagro Corporation. Submitted to WHO by Bayer AG, Leverkusen, Germany.

Nelson, D.L., White, R.G. Bauman, E.K., & Allen, A.D. (1968b). The effect of large injected doses of BAY Va 1470 to dogs. Unpublished report No. 23679 from Research and Development Department, Chemagro Corporation. Submitted to WHO by Bayer AG, Leverkusen, Germany.

Parton, J.W., Probst, G.S., & Garriott, M.L. (1988). The *in vivo* effects of 2,6-xylidine on induction of micronuclei in mouse bone marrow cells. *Mutat. Res.*, **206**, 281-283.

Parton, J.W., Beyers, J.E., Garriott, M.L., & Tamura, R.N. (1990). The evaluation of multiple dosing protocol for the mouse bone-marrow micronucleus assay using benzidine and 2,6-xylidine. *Mutat. Res.*, **234**, 165-168.

Poklis, A., MacKell, M.A., & Case, M.E.S. (1985). Xylazine in human tissues and fluids in a case of fatal drug abuse. *J. Anal. Toxicol.*, **9**, 234-236.

Ponder, S.W. & Clark, W.G. (1980). Prolonged depression of thermoregulation after xylazine administration to cats. *J. Vet. Pharmacol. Ther.*, **3**, 203-207.

Pütter, J. & Sagner, G. (1973). Chemical studies to detect residues of xylazine hydrochloride. *Vet. Med. Rev.*, **2**, 145-159.

Reynolds, S.M. (1994). Xylazine hydrochloride: teratology study in the rat. Unpublished report No. 94/BAG240/0268 from Pharmaco LSR. Submitted to WHO by Bayer AG, Leverkusen, Germany (GLP statement and report summary, only).

Robertson, S.A., Carter, S.W., Donovan, M., & Steele, C. (1990). Effects of intravenous xylazine hydrochloride on blood glucose, plasma insulin and rectal temperature in neonatal foals. *Equine Vet. J.*, **22**(1), 43-47.

Ruckebusch, Y. & Toutain, P.L. (1984). Specific antagonism of xylazine effects on reticulo-rumen motor function in cattle. *Vet. Med. Rev.*, **1**, 1-12.

Rudd, C.J., Mitchell, A.D., & Spalding, J. (1983). L5178Y Mouse lymphoma cell mutagenesis assay of coded chemicals incorporating analyses of the colony size distributions (Abstract No. Cd-19). *Environ. Mutagen.*, **5**(3), 419.

Samanta, A., Roffe, C., & Woods, K.L. (1990). Accidental self administration of xylazine in a veterinary nurse. *Postgrad. Med. J.*, **66**, 244-245.

Sagner, G. (1967). Bay Va 1470, anaesthetic, analgesic and sedative for veterinary medicine. Unpublished report No. 1470 from Bayer AG Institute of Pharmacology. Submitted to WHO by Bayer AG, Leverkusen, Germany.

Sagner, G., Hoffmeister, F., & Kroneberg, G. (1969). Pharmacological principles of a new preparation for analgesia, sedation and relaxation in veterinary medicine (Bay Ve 1470). *Vet. Med. Rev.*, **3**, 226-228.

Short, C.R., King, C., Sistrunk, P.W., & Kerr, K.M. (1983). Subacute toxicity of several ring-substituted dialkylanilines in the rat. *Fundam. Appl. Toxicol.*, **3**, 285-292.

Short, C.R., Joseph, M., & Hardy, M.L. (1989). Covalent binding of [^{14}C]-2,6-dimethylaniline to DNA of rat liver and ethmoid turbinate. *J. Toxicol. Environ. Health*, **27**, 85-89.

Singh, J., Peshin, P.K., Singh, A.P., & Nigam, J.M. (1983). Haemo-dynamic, acid base and blood gas alterations after xylazine administration in calves. *Indian J. Vet. Surg.*, **4**, 10-15.

Skarda, R.T., Jean, G., & Muir, W.W. (1990). Influence of tolazoline on caudal epidural administration of xylazine in cattle. *Am. J. Vet. Res.*, **51**(4), 556-560.

Spoerke, D.G., Hall, A.H., Grimes, M.J., Honea, B.N., & Rumack, B.H. (1986). Human overdose with veterinary tranquilizer xylazine. *Am. J. Emerg. Med.*, **4**, 222-224.

Swift, B.L. (1977). A technique for the surgical removal of the spleen in calves. *Vet. Med.-Small Anim. Clin.*, **January**, 77-79.

Symonds, W.H. (1976). The effect of xylazine on hepatic glucose production and blood flow rate in the lactating dairy cow. *Vet. Rec.*, **99**, 234-236.

Takase, I., Terada, H., & Fujii, T. (1976) Xylazine residues in organs and tissues of calves and milk of cows. Unpublished report No. 76/8278a from the Department of Veterinary Science, Faculty of Agriculture, Tokyo University of Agriculture and Technology. Submitted to WHO by Bayer AG, Leverkusen, Germany.

Tettenborn, D. (1969). BAY Va 1470. Subchronic toxicity for dogs during application as feed additive. Unpublished report No. 26692 (1731) from Bayer AG, Institute of Toxicology . Submitted to WHO by Bayer AG, Leverkusen, Germany.

Tettenborn, D. & Hobik, H.P. (1968a). BAY Va 1470. Chronic toxicity to rats from oral administration. Unpublished report No. 23414 (956) from Bayer AG, Institute of Pharmacology and Toxicology. Submitted to WHO by Bayer AG, Leverkusen, Germany.

Tettenborn, D. & Hobik, H.P. (1968b). BAY Va 1470. Subchronic toxicity to dogs from oral administration. Unpublished report No. 23415 (958) from Bayer AG, Institute of Pharmacology and Toxicology. Submitted to WHO by Bayer AG, Leverkusen, Germany.

Thurmon, J.C., Neff-Davis, C., Davis, L.E., Stoker, R.A., Benson, G.J., & Lock, T.F. (1982). Xylazine hydrochloride-induced hyperglycemia and hypoinsulinemia in thoroughbred horses. *J. Vet. Pharmacol. Ther.*, 5, 241-245.

Thurmon, J.C., Steffey, E.P., Zinkl, J.G., Waliner, M., & Howland, D. (1984). Xylazine causes transient dose-related hyperglycemia and increased urine volume in mares. *Am. J. Vet. Res.*, **45**(2), 224-227.

Trossmann, G. & Hobik, H.P. (1970). BAY Va 1470. Histopathological changes in organs of rats after feeding for 32 weeks. Unpublished report No. 28576 (956) from Bayer AG, Institute of Pathology and Histology. Submitted to WHO by Bayer AG, Leverkusen, Germany.

US National Toxicology Program (1990). Toxicology and Carcinogenesis Studies of 2,6-Xylidine (2,6-Dimethylaniline) (CAS No. 87-62-7) in Charles River CD Rats (Feed Studies). US National Toxicology Program, Research Triangle Park, NC, USA (NTP Technical Report No. 278; NIH Publication No. 90-2534).

Vernot, E.H., Macewen, J.D., Haun, C.C., & Kinkead, E.R. (1977). Acute toxicity and skin corrosion data for some organic and inorganic compounds and aqueous solutions. *Toxicol. Appl. Pharmacol.*, **42**, 417-423.

Wagner, A.E., Muir, W.W., & Hinchcliff, K.W. (1991). Cardio-vascular effects of xylazine and detomidine in horses. *Am. J. Vet. Res.*, **52**(5),:651-657.

Wasak, A. (1983). Haemetological and electrocardiographical changes in dogs after xylazine. *Med. Weter.*, **39**, 235-237.

Weiss, L.D., Generalovich, T., Heller, M.B., Paris, P.M., Stewart, R.D., Kaplan, R.M., & Thompson, D.R. (1987). Methemoglobin levels following intravenous lidocaine administration. *Ann. Emerg. Med.*, **16**(3), 323-325.

Zeiger, E., Anderson, B., Haworth, S., Lawlor, T., & Mortelmans, K. (1988). *Salmonella* mutagenicity tests: IV. Results from the testing of 300 chemicals. *Environ. Mol. Mutagen.*, **11**(12), 1-158.

ANTIMICROBIAL AGENTS

NEOMYCIN

First draft prepared by
Dr K.N. Woodward
Veterinary Medicines Directorate
Ministry of Agriculture, Fisheries and Food
Addlestone, Surrey, United Kingdom

1. EXPLANATION

Neomycin is an aminoglycoside antibiotic that was previously evaluated at the forty-third meeting of the Committee (Annex 1, reference 113). At that time the Committee established a temporary ADI of 0-30 μg/kg bw based on the NOEL of 6 mg/kg bw per day for ototoxicity in a 90-day study on the guinea-pig and a safety factor of 200. The ADI was made temporary in view of deficiencies in the genotoxicity data.

The *in vitro* genotoxicity data available at the forty-third meeting indicated that neomycin causes chromosomal aberrations, but only a limited number of studies were available and these had been poorly performed.

This monograph addendum summarizes the data that have become available since the previous evaluation.

2. BIOLOGICAL DATA

2.1 Toxicological studies

2.1.1 Special studies on genotoxicity

The results of the further genotoxicity studies with neomycin are summarized in Table 1.

Table 1. Results of genotoxicity studies on neomycin

Test system	Test object	Concen-tration	Results	Reference
Reverse mutation[1]	*Salmonella typhimurium* TA97A, TA98, TA100, TA1535	0.93–75 μg/plate	–	Mayo *et al.*, 1995
	Escherichia coli WP2 uvrA	0.93–75 μg/plate	–	Mayo *et al.*, 1995
Forward mutation[1]	Chinese hamster ovary cells, HGPRT locus	0–5000 μg/ml	–	Mayo & Aaron, 1995a
In vivo cyto-genetics test	mouse bone marrow	0–200/250 mg/kg bw per day	–	Mayo & Aaron, 1995b

[1] with and without rat liver metabolic activation (S9)

Neomycin was negative in the Ames *Salmonella typhimurium* reverse mutation assay, although the maximum concentration used was limited to 75 μg/plate, owing to cytotoxicity. Similarly, a negative result was obtained in a reversion assay with *Escherichia coli* WP2 uvrA, but again the maximum concentration was limited by cytotoxicity to 75 μg/plate. However, in a test for point mutations using Chinese hamster ovary cells (HGPRT locus), negative results were obtained with concentrations of up to 5000 μg/ml. A negative result was obtained in a cytogenetics test in mice where animals were given repeated intraperitoneal doses of up to 200 mg/kg bw per day. No studies were presented to show that neomycin distributes to bone marrow, but aminoglycosides are widely distributed in the body following absorption and it is highly likely, taking into account vascular perfusion of bone marrow, that the drug reached the target cells.

The data indicate that neomycin is not genotoxic.

3.   COMMENTS

The new data on genotoxicity considered by the Committee consisted of the results of a reverse mutation assay using *Salmonella typhimurium* and *Escherichia coli*, a forward mutation assay in Chinese hamster ovary cells, and an *in vivo* bone marrow cytogenetic assay in mice. All gave negative results.

The Committee concluded from these results that neomycin is not genotoxic.

4.   EVALUATION

The Committee established an ADI of 0-60 μg/kg bw, based on the NOEL of 6 mg/kg bw per day for ototoxicity in the guinea-pig and a safety factor of 100.

5.   REFERENCES

Mayo, J.K. & Aaron, C.S. (1995a). U-4567 (Neomycin Sulfate): Evaluation of U-4567 (Neomycin sulfate) in the ASA52/XPRT mammalian cell mutation assay with and without metabolic activation. Unpublished report No. 7228-95-125. Submitted to WHO by Pharmacia and Upjohn Company, Kalamazoo, MI, USA.

Mayo, J.K. & Aaron, C.S. (1995b). U-4567 (Neomycin sulfate): Evaluation of U-4567 (Neomycin sulfate) in the acute test for chemical induction of chromosome aberration in mouse bone marrow cells *in vivo*. Unpublished report No. 7228-95-130. Submitted to WHO by Pharmacia and Upjohn Company, Kalamazoo, MI, USA.

Mayo, J.K., Smith, A.L, & Aaron, C.S. (1995). Neomycin sulfate (U-4567): Evaluation of neomycin sulfate (U-4567) in the preincubation mutagenesis assay in bacteria (Ames Assay). Unpublished report No. 7228-94-133. Submitted to WHO by Pharmacia and Upjohn Company, Kalamazoo, MI, USA.

THIAMPHENICOL

First draft prepared by
Dr R. Fuchs,
Department of Experimental Toxicology and Ecotoxicology,
Institute for Medical Research and Occupational Health,
Zagreb, Croatia

1. EXPLANATION

Thiamphenicol is a broad-spectrum antimicrobial agent, structurally similar to chloramphenicol, used orally to control infections in humans, pigs, poultry and non-ruminating cattle. It is bacteriostatic for both Gram-positive and Gram-negative aerobes and for some anaerobes. It has not been previously evaluated by the Committee.

The molecular structure of thiamphenicol is shown below.

2. BIOLOGICAL DATA

2.1 Biochemical aspects

2.1.1 Absorption, distribution and excretion

2.1.1.1 Rats

After i.v. administration in rats the half-life of thiamphenicol was 46.3 minutes, compared to 21.5 minutes for chloramphenicol (Ferrari & Della Bella, 1974).

Pretreatment of rats with phenobarbital increased thiamphenicol half-life slightly from 46.3 to 55.2 minutes, whereas the chloramphenicol half-life was reduced from 21.5 minutes to 9.3 minutes (Della Bella et al., 1968a).

Following administration of 30 mg/kg to rats, thiamphenicol was eliminated in the urine almost entirely in the unchanged form: 62% following oral administration and 47% after i.m. administration (both at 48 hours). No significant glucuro-conjugation products were found. In the bile, 3.4% of the administered dose appeared unchanged and 10-12% appeared as conjugated products after 4 hours. Of the administered dose, 36% was eliminated in faeces after 75 hours, almost entirely as unchanged thiamphenicol. Distribution studies showed that levels in the kidney and liver were higher than in plasma, while brain concentrations were negligible (Gazzaniga, 1974).

2.1.1.2 Rabbits

The amount of metabolites recovered in urine and bile within 7 hours after i.v. administration was about 73% for thiamphenicol and 60%

for chloramphenicol, and the percentage of glucuronide to total amount recovered was 8% for thiamphenicol and 66% for chloramphenicol. The recovery of thiamphenicol into the bile was only 1%, mostly in unchanged form. About 3% of administered chloramphenicol was excreted into the bile and two-thirds of this consisted of metabolites (Uesugi *et al.*, 1974).

2.1.1.3 Dogs

Following intraduodenal administration of 70 mg/kg thiamphenicol or chloramphenicol, 30% of the thiamphenicol dose was found unchanged in urine within 8 hours, whereas only 8.3% of chloramphenicol was eliminated in the active form. After i.m. injection, the urinary elimination of active antibiotics within 8 hours was 24.2% for thiamphenicol but only 1.76% for chloramphenicol (Laplassote, 1962).

2.1.1.4 Pigs

To determine plasma and tissue concentrations of thiamphenicol in pigs following dietary treatment, 16 male pigs, approximately 7 weeks old, were fed thiamphenicol in the diet at a concentration of 900 mg/kg (equivalent to 30 mg/kg bw), twice daily for a period of 5 days. Three pigs were maintained as controls and fed basal diet only. Venous blood samples were taken prior to treatment and at various time-points during the study. The animals were killed at 4, 6, 8 and 10 days after treatment and samples of kidney, liver, muscle, fat and lung taken. Concentrations of thiamphenicol in plasma were measured following solvent extraction, using HPLC. On the second day of the dosing period, scouring and swelling or redness of the anus/perineal area were noted in all treated animals. The signs resolved within 4-10 days. The maximum mean plasma level of thiamphenicol (1.28 mg/litre) was demonstrated 8 hours after the first dose. Mean levels were in the range of 0.22-0.80 mg/litre during the dosing period and declined over the withdrawal period to give concentrations below or close to the limit of detection (0.01-0.08 mg/litre) from 4 hours to 5 days after the end of treatment. The results of the analysis of tissue samples were not reported (Redgrave *et al.*, 1991).

2.1.1.5 Humans

After a 500 mg oral dose of each of thiamphenicol and chloramphenicol, the plasma levels appear to be similar. A high level of active thiamphenicol and a low level of chloramphenicol were found in the urine (53.1% and 9.2%, respectively, after 24 hours). The half-life of thiamphenicol is significantly increased in renal insufficiency, but is almost unaffected by liver cirrhosis (Azzolini *et al.*, 1972).

2.2   Toxicological studies

2.2.1   Acute toxicity studies

The acute toxicity of thiamphenicol and thiamphenicol-glycinate is given in Table 1.

Table 1. Acute toxicity of thiamphenicol (TAP) and thiamphenicol-glycinate (TAP-G)

Species	Sex	Route	LD_{50} (mg/kg bw)		Reference
			TAP	TAP-G	
Mouse	M & F	oral	> 5000		Bonanomi, 1978
	M & F	i.p.	> 3000	1550	Bonanomi, 1978
	M & F	i.v.		450	Bonanomi, 1978
Rat	M & F	oral	> 5000		Bonanomi, 1978
	M & F	i.p.	> 3000	1750	Bonanomi, 1978
	M & F	i.v.		470	Bonanomi, 1978

The predominant clinical signs in the animal species tested were sedation and piloerection after oral dosing, and dyspnoea, cyanosis, motility disorders and respiratory arrest after parenteral administration.

2.2.2   Short-term toxicity studies

2.2.2.1   Rats

The oral toxicity of thiamphenicol, when administered as an aqueous suspension in 0.5% methocel for 13 week, has been investigated in four groups of rats (Sprague-Dawley 30/sex/group). Thiamphenicol was administered by oral gavage at dose levels of 30, 45, 65 or 100 mg/kg bw per day, while a fifth control group received 0.5% methocel only. The first 15 animals of each group were killed on completion of the treatment period, and the remaining 15 after a recovery period of 8 weeks. Mortality during the treatment period was elevated among animals of both sexes receiving 100 mg/kg bw per day, and during the recovery period mortality was similar in all groups. At levels of 65 and 100 mg/kg bw per day, pallor, hair loss, prostration, hunched posture and flaccid musculature were seen.

After the recovery phase the incidence and the severity of symptoms were similar in all groups including controls. Body weight stasis or loss was seen in animals receiving 65 and 100 mg/kg bw per day, and after 8 weeks recovery body weight was similar in all groups. Food intake was reduced at dose levels of 45, 65 and 100 mg/kg bw per day, with immediate improvement after cessation of treatment.

In all treated animals there were changes in erythrocyte parameters, differential and total leukocyte counts and clotting parameters, all of which were dose-related. After the recovery period erythrocyte and leukocyte counts were still low in males treated with 65 and 100 mg/kg bw per day. Plasma levels of urea, triglycerides and total protein were altered from week 7 in animals treated with 45 mg/kg bw per day and after the end of treatment also in males receiving 30 mg/kg bw per day. Parameters associated with liver and kidney function were affected at the two higher doses. After 8 weeks full recovery was considered to have occurred. The weights of all the main organs were reduced at dose levels of 65 and 100 mg/kg bw per day, but after the recovery period only the testis weight was still reduced. Postmortem examination revealed effects on the gastrointestinal tract and spleen in both sexes and in the liver, thymus and testis of males at the highest dose level only. After 8 weeks only the testis weights were still reduced. The erythroid/myeloid cell ratio was increased in both sexes at doses of 65 or 100 mg/kg bw per day, and after 8 weeks was still slightly higher than usual.

Treatment-related changes were seen in tissues with high cell turnover rates; most of them recovered after a period of respite from treatment. Animals in the groups treated with 30 and 45 mg/kg per day showed no histopathological findings except hepatocytic reduced basophilia (male, 45 mg/kg bw per day) and increased splenic extramedullary haematopoiesis (female 45 mg/kg bw per day). Testicular germinal epithelial deficit was present at doses above 45 mg/kg, and caecal oedema and adnexal atrophy were present after the recovery period at the highest dose level. The NOEL was 30 mg/kg bw per day (Marubini et al., 1991).

In a 6-month toxicity study in the rat, thiamphenicol was administered as an aqueous suspension in 2% gum arabic by stomach tube 6 days/week to 180 Wistar rats (30/sex/group) at dose levels of 0, 40 or 120 mg/kg bw per day. Body weight and food intake were recorded twice weekly in the first 4 weeks of treatment, and thereafter only once a week. At the end of weeks 4, 8, 16 and 24, ten animals from each group were killed following collection of 2-hour urine sample. Haematology, clinical chemistry assays and urinanalysis were performed on all tested animals. Histological examinations were carried out on the lungs, spermatozoa, blood and bone marrow smears and samples of main organs. There was a decrease in food intake at a dose level of 120 mg/kg bw per day and a dose- and time-related decrease in body weight gain in females. No effect was observed on erythropoiesis or on hepatic or renal function. Urinanalysis showed the presence of albumin and haemoglobin in the high-dose rats. No gross pathological variations in organ weights were observed, but irritative changes in the gastrointestinal mucosa and high incidence of monolateral spontaneous hydronephrosis were seen both in treated and control animals. Histopathological examination performed on control and high-dose rats revealed a slight effect on the morphology of spermatozoa in the high-dose group at the 8th, 16th and 24th week of treatment (Della Bella *et al.*, 1968b).

2.2.2.2 Rabbits

Thiamphenicol-glycinate was administered subcutaneously to groups of rabbits ("Fauve de Bourgogne" seven/sex/group) at dose levels of 0, 25, 50 or 100 mg/kg bw per day, 6 days/week for 12 weeks. Animals were weighed weekly and submitted to haematological (erythrocyte and leukocyte counts with differentials) and biochemical (urea, reducing sugars, chlorides) examinations before treatment, after 6 weeks and at the end of the treatment period. After macroscopic examination of the viscera, the various organs were examined histologically. During the treatment two control animals and three in the high-dose group died (no autopsies were carried out). A reduction in polymorphonuclear neutrophils in male rabbits in all treated groups and a slight fall in erythrocyte levels in high-dose females were observed. Anatomical and pathological examination provided no evidence of damage attributable to thiamphenicol-glycinate administration (Brunaud, 1965).

2.2.2.3 Dogs

In a 7-week study in beagle dogs (four/sex/group), thiamphenicol was administered orally in gelatin capsules at dose levels of 0, 40 or 80 mg/kg bw per day. Two males and two female dogs were killed at the end of the 7-week treatment, and the remaining animals were kept for further 12 weeks without treatment before being killed. Behaviour and body weights were recorded, haematology and clinical chemistry examinations and urinanalysis were conducted on all animals pretest and at various intervals during the study. Complete gross postmortem examination, organ weighing and histopathological evaluation were conducted on all animals. The animals in the two treated groups developed diarrhoea soon after the beginning of treatment, which spontaneously regressed in the low-dose group but persisted in the high-dose group. At 80 mg/kg bw per day reduced food consumption with loss of weight and muscular asthenia occurred, and vomiting was observed in some cases. Four of these dogs were killed at the end of the 4th week. Slight loss of weight was observed in two dogs in the low-dose group.

Decreases in haematocrit, haemoglobin concentration and erythrocyte count were seen in both treated groups, but did not appear to be dose-related, and returned to normal on withdrawal of treatment. Slight increase in proteinuria was observed in the last few weeks of treatment. At 40 mg/kg bw per day superficial erosion of the gall bladder mucosa and at 80 mg/kg bw per day haemorrhagic ulcers in the gall bladder, diffuse muco-membranous enteritis and thymic involution were seen in dogs killed at the end of treatment period. In dogs killed after the 12-week recovery period no differences were observed between treated and control animals. Histological examination after 7 weeks revealed in the high-dose group severe cholecystitis, chronic sclerosing pancreatitis, enteritis, severe depletion of haematopoietic marrow and lymphoid thymus depletion. Two dogs in the low-dose group showed depletion of germinal epithelium in the testes and multinucleated cells in seminiferous tubules, which were not seen in the high-dose group. None of the changes observed at 7 weeks were detectable in the animals kept for 12 weeks after cessation of treatment (Bonanomi et al., 1978).

Thiamphenicol was administered orally in gelatin capsules to 24 beagle dogs (four/sex/group) at dose levels of 30, 60 or 120 mg/kg bw per day for a period of 4 weeks. Physical observations, ophthalmoscopic examinations, and body weight and food consumption measurements were performed before treatment and over selected intervals during the treatment. Haematology, clinical chemistry and urinanalysis were conducted on all animals pretest and at study termination. Complete gross postmortem examinations, organ weight and histopathological evaluation were conducted on all animals. The body weights of the high-dose animals were slightly lower than those of controls at week 3 in both sexes, and at week 4 in males only. At 60 and 120 mg/kg bw per day absolute and relative liver weights in male dogs were greater than in controls, and relative liver weights were also increased in females. Microscopically, hepatocellular hypertrophy was present in the liver of mid- and high-dose animals, which correlated with the increase in liver weights in these groups. No other parameter evaluated showed evidence of adverse treatment-related effects. The NOEL was 30 mg/kg bw per day (Kelly & Daly, 1990).

Thiamphenicol was administered orally in gelatin capsules to 56 beagle dogs (seven/sex/group) at dose levels of 15, 30 or 60 mg/kg bw per day. After 6 months of treatment four animals/sex/group were sacrificed, and the remaining three animals/sex/group were kept for a 2-month recovery period. Physical observations, ophthalmoscopic examinations, body weight, food consumption, haematology and clinical chemistry examinations and urinanalysis were conducted on all animals pretest and on all surviving animals at selected intervals during the treatment and recovery period. Complete gross postmortem examinations, organ weight and histopathological evaluation were conducted on all animals. One control male was found dead during the study. One male and one female dog in the highest dose group were moribund and had to be killed. Clinical signs prior to death included lethargy, poor food consumption, emaciation, tremors and dehydration. Physical findings related to thiamphenicol administration were tremors, lethargy, irregular gait and excessive licking or chewing in the high-dose group. Tremors were also present in the mid-dose animals. These signs were seen during the last two months of the study and were not present at the end of the recovery period. Body weights of the high-dose males during the study were 4 to 18% lower than those of controls. Decreased erythrocyte counts and mean haematocrit values were seen in high-dose males at weeks 6 and 13 and at termination of the study, and in high- and mid-dose females at week 13 and at termination of the study. After the recovery period no differences were observed in the haematological parameters between control and treated animals. No treatment-related effects were seen in the bone marrow smear examinations. Mean serum cholesterol and phospholipid levels of the mid- and high-dose males at week 6 and 13 and at termination of the study were greater than control values. The same parameters were elevated in high-dose females at the end of the study. Mean serum glucose levels of males at 60 mg/kg bw per day and females at 30 and 60 mg/kg bw per day were significantly increased. Mean fibrinogen values of high-dose females were elevated at the end of the study. Relative liver weights were increased at mid- and high-dose levels. Pathological lesions related to treatment were seen in sections of the thymus (exacerbation of involution), bone marrow (decreased cellularity), liver (centrilobular necrosis and pigment deposition), testes (focal and diffuse tubular atrophy) and oesophagus (ulceration) from high-dose animals of both sexes, mostly occurring in animals killed in a moribund condition. No alterations were noted in any of the tissues examined microscopically from animals that were allowed to recover after treatment. The NOEL was 15 mg/kg bw per day (Kelly & Daly, 1991).

2.2.2.4   Pigs

A study designed to determine tolerance in pigs to treatment with thiamphenicol at three times the recommended dose for 5 days and at the normal recommended dose for 15 days was conducted with 16 weaned large white hybrid pigs (two pigs/sex/group). Thiamphenicol was administered in the diet at dose levels of 30 or 90 mg/kg bw per day for 5 days or 30 mg/kg bw per day for 15 days. The control group of animals was fed basal diet only. Clinical signs, body weight and food consumption were recorded. Blood, urine and faecal samples were obtained before dosing and at various time-points during the study. No significant treatment-related clinical abnormalities were noted. Body weight changes and food consumption were within normal limits. No consistent treatment-related differences in haematological and biochemical parameters or in urinanalysis values were observed. The authors concluded that treatment with thiamphenicol had no significant adverse effects on general health, body weight, food consumption or standard clinical pathology parameters (Roberts *et al.*, 1989).

In a 4-week toxicity study, groups of pigs (Large White hybrid, four/sex/group) were fed 25, 50 or 100 mg thiamphenicol/kg bw per day. The control group of animals was fed basal diet only. Clinical signs, body weight and food consumption were recorded. Blood, urine and faecal samples were obtained before dosing and during week 4 for clinical pathological investigations. At the end of the 4-week dosing period, pigs were killed, and selected tissues were processed for histological examination. In all groups treated with thiamphenicol, swelling and erythema of the anus, vulva/testes and perineal area, tail and hocks were observed on the second day of the dosing period. These effects were consistent with scouring and irritancy, e.g., as a result of disruption of normal gastrointestinal flora activity, and disappeared within 1 to 13 days. All pigs remained in good health thereafter. Slight reductions in body weight gain and food consumption were noted at 50 and 100 mg/kg bw per day. At week 4 there was a slight reduction in mean PCV, haemoglobin concentration and erythrocyte counts in animals receiving 100 mg/kg bw per day and a treatment-related reduction of urinary pH in all groups dosed with thiamphenicol. At the end of the study, increases in liver and kidney weights in pigs fed 50 and 100 mg/kg bw per day were observed. On histological examination, treatment-related changes were found in the highest dose group only: an increase in vacuolation and fat in renal tubular epithelium and, in some animals, minimal diffuse hepatocyte vacuolation and hepatocyte fat (Cameron *et al.*, 1990).

2.2.3   Long-term toxicity/carcinogenicity study

2.2.3.1   Rats

As a range-finding study for the dose selection in a 2-year carcinogenicity study, four groups of F-344 rats (10/sex/group) were given drinking-water containing 0, 125, 250 or 500 mg/litre thiamphenicol (equal to 9, 17 or 36 mg/kg bw per day for males and 12, 29 or 39 mg/kg bw per day for females) for 13 weeks. The examinations at the end of the study covered clinical observations, water consumption, body weight changes, haematological parameters, serum biochemistry, organ weights and gross and microscopic appearance. In haematological examinations anaemic changes were observed in males treated with 250 mg/litre or more and high-dose females. Similar changes, such as increased MCV and increased related counts, were observed in males of the 125 mg/litre group and females of the 250 and 125 mg/litre groups. At autopsy, enlargement of the caecum was seen in treated groups of both sexes. Histologically, the highest dose animals showed decreased haematopoiesis of the bone marrow, decreased spermatogenesis of the testis and sperm granulomas of the epididymis. Sperm granulomas in the epididymis were also seen in some of the animals in the 250 mg/litre group. Based on the results of this pilot study, a 2-year carcinogenicity study of thiamphenicol was performed in rats. Three groups of F-344 rats (50/sex/group) were given drinking-water containing 0, 125 or 250 mg thiamphenicol/litre (equal to 8 or 16 mg/kg bw per day for males and 9.7 or 19 mg/kg bw per day for females) for 104 weeks. All surviving animals subjected to 4-week withdrawal of the test chemical after the end of the treatment were killed for full histopathological examinations. The high-dose animals showed decreased body weight gain, but the incidence of tumours in treated groups was not significantly higher than that of controls (Maekawa, 1996; summary report only was available).

2.2.4 Reproductive toxicity studies

2.2.4.1 Rats

In a fertility study on male Wistar rats, oral treatment with thiamphenicol for 2 or 3 months (30 animals per dose level, 10 per treatment time), at dose levels of 120, 180 or 240 mg/kg bw per day, resulted in reduction in the number of tubular germinal cells, which was more marked at the highest dose level. Ten animals of each group were treated for 4 weeks, ten for 8 weeks and the last ten for 12 weeks. At the end of each test period 5 animals of each dose group were killed and necropsied, while the remaining 5 rats were mated with normal females. At 240 mg/kg bw per day extensive testicular hypotrophy, together with severe depletion of the germinal epithelium 21 days after withdrawal of treatment, was seen. Histological changes coincided with a reduction of the fertility index, which gradually recovered within 50 days. Litters from matings between treated males and normal females were normal in number and weight, and no morphological abnormalities were observed. The concentration ratio of thiamphenicol between testes and plasma after administration of 240 mg/kg bw per day thiamphenicol was 1, indicating the absence of accumulation in testes (Della Bella *et al.*, 1967).

Groups of 21 Sprague-Dawley rats were given thiamphenicol orally (30, 60 or 120 mg/kg bw per day daily) from day 15 of gestation to day 21 postpartum. In groups receiving 60 and 120 mg/kg bw per day a higher post-implantation loss, slight weight reduction at birth and increased rate of perinatal mortality were observed. No malformations were observed. Development of pups was inhibited during the lactation period with a consistent dose-dependent relationship. From day 30 postpartum a good recovery was observed in all groups. Sexual behaviour and fertility of F_1 animals were normal, and the F_2 generation showed no signs of abnormal development (Bonanomi *et al.*, 1980).

2.2.5   Special studies on embryotoxicity and teratogenicity

2.2.5.1   Rats

Teratogenicity studies were carried out on 195 mature female Wistar rats (15/group) given thiamphenicol orally at dose levels of 40, 80 or 160 mg/kg bw per day from days 1 to 21 of pregnancy and 80 or 960 mg/kg bw per day from days 1 to 7, 7 to 14 or 14 to 21 of gestation. Thiamphenicol did not induce any teratogenic effects in any of the four studies carried out. When the treatment period was 1-21 days, a dose-related increase in resorptions was noted, and newborns had a high mortality rate in the second and third week of life, particularly in the 40 mg/kg bw per day group. In rats treated on days 1-7 of gestation, a non-dose-related increase in resorptions and an increased mortality in newborns in the third week after birth were observed. When the treatment period was from 7 to 14 days, complete resorption of fetuses occurred at 160 mg/kg bw per day. The mean number of newborns per litter was reduced at 80 mg/kg bw per day and there was a high mortality of newborns in the first week. An increased mortality among newborns of the group treated with 80 mg/kg bw per day during days 14-21 of gestation was observed (Bonanomi & De Paoli, 1969).

When the inhibition of mitochondrial functions induced by thiamphenicol was compared with the inhibition of overall embryonic development, it appeared that mitochondrial respiration was the rate-limiting step for the embryotoxic effects of thiamphenicol. Because of the lack of specificity of these effects, prenatal mortality rather than teratogenic effects was seen (Bass *et al.*, 1978).

2.2.5.2   Rabbits

A teratogenicity study was performed with 50 New Zealand white rabbits administered thiamphenicol orally at dose levels of 5, 30, 60 or 80 mg/kg bw per day from the 8th to the 16th day of pregnancy. The highest dose resulted in a complete resorption of implantation due to the toxic effects on the mothers. Data obtained in all treated groups showed moderate fetal toxicity with a dose-related increase in abortion rate and resorption. No skeletal malformations were found in fetuses (Bonanomi *et al.*, 1974).

Thiamphenicol in 0.5% Methocel K15M was administered daily by oral gavage to female rabbits (16/group) from day 6 to day 18 of gestation at doses of 1.25, 2.5 or 5.0 mg/kg bw per day. Control animals received the vehicle alone. The females were killed on gestation day 29 and subjected to postmortem examination. All fetuses were examined for external and internal abnormalities and skeletal changes. No clinical signs attributable to treatment were observed. Mild signs of maternal toxicity were observed in mid- and high-dose animals, indicated by reduction in body weight changes during the treatment period.

Necropsy findings in females on gestation day 29 were incidental, with no relation to treatment. Litter parameters and sex ratios did not show any significant difference between groups. Mean fetal weight was decreased in the high-dose group. Two fetuses in the mid-dose group and one in the control group were malformed, with cleft palate, abnormally shaped head, extra digits and incomplete flexure of the hind limbs. The number of small fetuses in the high-dose group was higher than in the control group. The few anomalies seen during the internal examination were not considered to be treatment-related. Skeletal examination revealed no differences between high-dose fetuses and controls. The authors concluded that thiamphenicol administered by oral gavage at concentrations of 1.25, 2.5 and 5 mg/kg per day had no effects on pregnancy or embryo-fetal development. However, the Committee concluded that the NOEL for embryotoxicity under the conditions of this experiment was 1.25 mg/kg per day (Sisti, 1994).

2.2.6   Special studies on genotoxicity

The results of genotoxicity assays on thiamphenicol are given in Table 2.

2.2.7   Special studies on immune responses

The effects on spontaneous nephritis in NZB x OUW hybrid mice (32-39 males/group) were investigated in a study involving lifetime administration of thiamphenicol in feed at dose levels of 25, 50 and 250 mg/kg bw per day. Body weight, urinary protein, limited haematology, organ weights and histopathology investigations were reported. The prolonged treatment with thiamphenicol at dose rates of ≥ 50 mg/kg reduced the severity of the spontaneous renal disease and significantly extended lifespan, compared to untreated controls. The immunosuppressive effect of thiamphenicol was demonstrated histologically by a reduction in immune-complex deposition in the glomeruli. No evidence of malignancy or premalignant signs was seen (Simpson *et al.*, 1979).

2.2.8   Special studies on microbiological effects

2.2.8.1   *In vivo*

In a study of thiamphenicol-induced changes in mouse intestinal microflora, 100 female albino mice were divided into 10 subgroups, and five of these groups were treated with thiamphenicol at concentrations of 40 µg/kg diet for 35 days. Samples of intestinal content were taken from the

Table 2. Genotoxicity assays on thiamphenicol

Test system	Test object	Concentration	Results	Reference
Ames test[1]	*S. typhimurium* TA98, TA100, TA1535, TA1537, TA1538	0.5-50 μg/plate[2]	Negative	Pinasi *et al.*, 1990a
Gene conversion and mitotic crossing over[1]	*Saccharomyces cerevisiae*	2.8-140.3 mM[3]	Negative	Marca & Bonanomi, 1979
Gene mutation assay[1]	Chinese hamster V79 cells	50-5000 μg/ml[4]	Negative	Pinasi *et al.*, 1990b
Chromosomal aberrations[1]	Cultured human lymphocytes	700-3250 μg/ml[5]	Negative	Mosesso & Driedger, 1989
DNA repair test	Primary rat hepatocytes	500 and 1000 μg/kg[6]	Negative	Bichet, 1985
In vivo micronucleus assay	Mouse bone marrow	2500 and 5000 mg/kg[7]	Negative	Pinasi *et al.*, 1990c

[1] Both with and without rat liver S9 fraction
[2] Methyl methanesulfonate and cyclophosphamide were used as positive controls
[3] 2-Nitrofluorene, 9-aminoacridine, sodium azide and 2-aminoanthracene were used as positive controls
[4] Ethyl ethanesulfonate and *N*-dimethylnitrosoamine were used as positive controls
[5] Mitomycin-C and cyclophosphamide were used as positive controls
[6] 2-Aminofluorene was used as positive control
[7] Cyclophosphamide was used as positive control

caecum for bacteriological analysis before the treatment and after 7, 14, 28 and 35 days. Microorganisms were isolated by preparing serial dilutions of intestinal content, and the most representative bacteria of the mouse intestinal microflora were cultured on specific media. Their sensitivity to thiamphenicol was assessed by calculating the minimal inhibitory concentration (MIC) of the drug. The results show that the mean counts of various microorganisms did not differ significantly between control mice and those fed with thiamphenicol. The numbers of the bacterial populations did vary at the different sampling times, and in some cases the difference from pre-treatment results was significant, but confidence limits were the same in treated and control mice killed at the same time. The investigation shows that there were no appreciable differences in the type or amount of bacterial flora related to thiamphenicol administration. Differences in the distribution of genera such as *Diplococcus* sp. and *Escherichia* sp. were similar to those in controls. The thiamphenicol MIC values for the numerous strains tested indicate that addition to feed at concentrations corresponding to the proposed MRL of 40 μg/kg feed does not select for drug-resistant strains and has no effect on the qualitative or quantitative composition of the intestinal microflora. The MIC remained unchanged throughout the 35 days of the study (Poli, 1994).

2.2.8.2   *In vitro*

In vitro antibacterial activity of thiamphenicol against 489 bacterial isolates from infected animals was determined by the agar dilution method. In the case of mycoplasmas, however, MICs were determined by the broth dilution method. Depending on bacterial strains tested estimations were made under aerobic or anaerobic conditions. The results are presented in Table 3 (Albini, 1989).

Table 3. Antibacterial activity of thiamphenicol against 489 animal pathogens

Organisms	Number of isolates	MIC (μg/ml)		Range
		MIC_{50}	MIC_{90}	
Bacteroides spp.	11	2	16	1-128
Bordetella spp.	9	32	32	16-32
Campylobacter spp.	17	8	16	4-16
Clostridium spp.	37	2	4	0.25-16
Corynebacterium spp.	10	2	16	2-16
Escherichia coli	61	128	>128	16 - >128
Haemophilus pleuropneumoniae	7	0.5	1	0.5-1
Micrococcus spp.	6	0.5	0.5	0.5-8
Mycoplasma spp.	9	1	2	0.125-4
Pasteurella spp.	71	1	2	0.25-128
Salmonella spp.	34	32	32	8 - >128
Staphylococcus spp.	94	8	32	4 - >128
Streptococcus spp.	123	2	4	0.5 - >128

Data on the sensitivity of normal components of human intestinal microflora are presented in Table 4 (Sutter & Finegold, 1976; Schioppacassi, 1992).

Table 4. Antibacterial activity of thiamphenicol against 261 strains of anaerobic bacteria isolated from humans (From: Sutter & Finegold, 1976)

Bacteria	No. of strains tested	Cumulative % susceptible to indicated concentration ($\mu g/ml$)									
		≤ 0.1	0.5	1.0	2.0	4.0	8.0	16.0	32.0	64.0	128.0
Bacteroides fragilis	42				5	21	71	100			
Bacteroides melaninogenicus	59	9	24	51	90	100					
Other *Bacteroides* and *Selenomonas*	21		19	24	57	76	86	91	95		100
Fusobacterium nucleatum	8		100								
Other *Fusobacterium*	12		42	92	100						
Peptococcus and *Gaffkya*	17			35	71	100					
Peptostreptococcus	15		33	47	93	100					
Anaerobic and microaerophilic streptococci	6				50	83		100			
Gram-negative cocci	7		43	86	100						
Eubacterium	7				43	57	100				
Arachnia propionica	2			50	100						
Propionibacterium	4		50	75			100				
Actinomyces	16		25	56	94			100			
Lactobacillus	10		10	30	50	90			100		
Clostridium perfringens	8					100					
Other *Clostridium*	27			4	22	63	78	96		100	

2.3 Observations in humans

Reversible dose-related bone marrow suppression is seen after thiamphenicol treatment and is attributed to its inhibitory effect on mitochondrial protein synthesis (Nijhof & Kroon, 1974). Reversible inhibition of myeloid and erythroid colony growth by thiamphenicol, resulting from an inhibition of mitochondrial protein synthesis, is consistent with the reversible bone marrow suppression induced by this drug (Yunis & Gross, 1975).

A study of clinical reports on the use of thiamphenicol in 16 631 cases from 1968 to 1977 in Japan revealed that blood disorders occurred in 41 (0.46%) out of 8848 patients receiving thiamphenicol glycinate therapy and 28 (0.36%) out of 7783 patients receiving thiamphenicol. The disorders were dose-dependent, occurring mainly in erythrocytes, and disappeared spontaneously on discontinuation of the drug (Tomoeda & Yamamoto, 1981).

The *para*-nitro group of chloramphenicol has been shown to have a central role in the pathogenesis of aplastic anaemia, probably as a result of its reduction to the highly toxic nitroso metabolite, which is a potent inhibitor of DNA synthesis. Thiamphenicol does not show this activity and the absence of the *para*-nitro group is therefore advanced as evidence that thiamphenicol cannot induce aplastic anaemia (Murray *et al.*, 1983).

Thiamphenicol has been used extensively in human medicine for over 25 years. Total human exposure to thiamphenicol up to 1987 has been estimated at 130-650 million exposures, assuming an average course of therapy of 7.5-15 g (Personal communication from Dr S. Biressi, Zambon Group SpA, Italy to Dr R.D. Agostino, Farmaquest Co.; submitted to WHO by Zambon Group SpA, Italy).

Epidemiological studies have not established any causal association between thiamphenicol treatment and irreversible aplastic anaemia (TAP Pharmaceuticals Inc., 1987). Statistical analysis of data obtained in these studies support the argument that the risk of aplastic anaemia from exposure to thiamphenicol is similar to the background risk of idiopathic aplastic anaemia (from 1 in 200 000 to 1 in 800 000) (Kelly & Kaufman, 1989).

3.   COMMENTS

Studies on thiamphenicol available for evaluation included data on pharmacokinetics, acute toxicity, short-term toxicity, reproductive toxicity, developmental toxicology, genotoxicity, and limited information on long-term toxicity. Studies on microbiological effects of thiamphenicol and epidemiological data on humans were also considered by the Committee. The Committee noted that many of the studies were conducted utilizing protocols that would not meet contemporary standards, and therefore the substance was evaluated under the procedures developed for drugs with a long history of use (Annex 1, reference 104).

The pharmacokinetic data showed that the drug is rapidly absorbed when administered by oral or parenteral routes. After intravenous administration in rats, the half-life was estimated to be 46 minutes. The main route of excretion in humans and animals is in the urine; approximately 60% of an oral dose of 30 mg/kg bw was excreted unchanged in the urine over a 24-hour period.

Single oral doses of thiamphenicol were of low toxicity to mice and rats (LD_{50} > 3000 mg/kg bw).

Short-term oral toxicity studies with thiamphenicol were performed in rats, dogs and pigs, the results are described in the following paragraphs.

In a 13-week study in rats at dose levels of 30, 45, 65 or 100 mg/kg bw per day, increased mortality was observed among animals given 100 mg/kg bw per day. In a 6-month study in rats, where the highest dose used was 120 mg/kg bw per day, increased mortality was not reported. In both studies, decrease in body weight gain during treatment occurred at doses of ≥ 65 mg/kg bw per day. Dose-related decreases in red blood cell parameters, differential and total white blood cell counts, and clotting parameters were observed in the 13-week study, but the same effects were not reported in the 6-month study. Testicular germinal epithelial cell depletion was seen at doses above 45 mg/kg bw per day in the 13-week study, and a dose of 30 mg/kg bw per day was considered to be the NOEL.

Dogs were given 40 or 80 mg thiamphenicol/kg bw per day for 7 weeks. At both dose levels decreases in body weight were observed, as well as reversible decreases in haematocrit, haemoglobin concentration and erythrocyte count. At 40 mg/kg bw per day, superficial erosion of the gall bladder mucosa was observed. The higher dose level resulted in haemorrhagic ulcers in the gall bladder, diffuse mucomembranous enteritis and early thymic involution. Two dogs in the low-dose group had testicular germinal epithelial cell depletion and multinucleated cells in the seminiferous tubules.

When thiamphenicol was given to dogs at doses of 30, 60 or 120 mg/kg bw per day for 4 weeks, the body weights of the high-dose animals were slightly lower than those of controls. In the mid- and high-dose groups, increases in absolute and relative liver weights were observed. Hepato-cellular hypertrophy was present in the liver of dogs given 60 or 120 mg/kg bw per day.

In a 6-month study dogs were given thiamphenicol at doses of 15, 30 or 60 mg/kg bw per day. The body weights of high-dose males during the study were up to 18% lower than those of controls. The main haemato-logical findings were decreases in red blood cell parameters at the highest dose level. Increases were noted in mean serum cholesterol level and phospholipid concentrations in males (30 and 60 mg/kg bw per day groups) and females (60 mg/kg bw per day group), and in the mean serum glucose concentration of males (60 mg/kg bw per day group) and females (30 and 60 mg/kg bw per day groups). The relative liver weights at the mid- and high-dose levels were increased. Histopathological lesions related to treatment were seen in the thymus (early involution), bone marrow (decreased cellularity), testes (focal and diffuse tubular atrophy) and oesophagus (ulceration) of high-dose animals. The NOEL was 15 mg/kg bw per day.

In a 4-week study, pigs were treated with 25, 50 or 100 mg thiamphenicol/kg bw per day. In the highest-dose group, slight reduction in body weight gain, as well as reductions in mean packed cell volume, haemoglobin concentration and erythrocyte counts, were observed, and histological examination showed vacuolation in renal tubular epithelial cells and mild diffuse hepatocyte vacuolation. In all treated groups treatment-related reduction in urine pH was observed.

A summary report of a two-year carcinogenicity study in rats, including a range-finding study, was available to the Committee. Rats were given 125 or 250 mg/kg thiamphenicol in drinking-water (equal to 8 or 16 mg/kg bw per day for males and 10 or 19 mg/kg bw per day for females) for 104 weeks. The highest-dose animals showed a decrease in body weight gain, but there was no significant increase in the incidence of tumours in treated groups compared to control animals.

In a long-term study in mice (32-39 males/group), which was designed primarily to investigate the effects of thiamphenicol on immune responses, thiamphenicol was administered orally at doses of 25, 50 or 250 mg/kg bw per day. No evidence of neoplastic or preneoplastic changes was observed.

In a study to determine the effect of thiamphenicol on fertility in rats, the drug was administered orally at doses of 120, 180 or 240 mg/kg bw per day for 2 or 3 months. Thirty male rats were used per dose level (10 per treatment period). From each treatment-period group, half of the animals were killed for histopathological examination at the given time intervals, while the remaining males were mated with untreated females. Reductions in the number of germinal epithelial cells in testes of all treated animals were observed. These changes were present up to 21 days after termination of treatment, and full recovery was observed by 50 days. Histological changes correlated with the fertility index. Litters from matings between treated males and non-treated females were normal in number and no physical abnormalities were reported.

Thiamphenicol was given orally to rats from day 15 of gestation to day 21 post-partum at doses of 30, 60 or 120 mg/kg bw per day. In the mid- and high-dose groups, there were higher post-implantation losses and increased rates of perinatal mortality. Physical development of pups was inhibited during the lactation period in a dose-dependent manner. Sexual behaviour and fertility of F_1 animals were normal, and animals in the F_2 generation showed no abnormalities.

In four teratogenicity studies in rats, thiamphenicol was administered orally at dosages of 40, 80 or 160 mg/kg bw per day from days 1 to 21 of pregnancy or of 80 or 960 mg/kg bw per day over critical days of gestation (either 1-7, 1-21, 7-14 or 14-21). No teratogenic effects were observed. In all animals treated from days 1 to 21, a dose-related increase in resorption was noted and newborn pups had an elevated mortality rate.

A teratogenicity study was performed in rabbits using oral doses of 5, 30, 60 or 80 mg/kg bw per day from days 8 to 16 of gestation. Complete resorption of embryos occurred at 80 mg/kg bw per day. In other treated groups, moderate fetal toxicity and dose-related increases in abortion rate and resorption were reported. No malformations were found in fetuses.

In another teratogenicity study, rabbits received oral doses of thiamphenicol at doses of 1.25, 2.5 or 5 mg/kg bw per day from days 6 to 18 of gestation. Mild maternal toxicity was observed in mid- and high-dose animals in the form of depressed body weights during the treatment. No effects were observed on embryo-fetal development. The NOEL was 1.25 mg/kg bw per day.

Thiamphenicol gave negative results in five *in vitro* genotoxicity tests and in an *in vivo* micronucleus assay using mouse bone marrow.

The Committee considered data from human epidemiological studies and concluded that there was no evidence that thiamphenicol can induce aplastic anaemia, in contrast to the structurally related compound, chloramphenicol.

The Committee considered data from several *in vitro* studies on the minimum inhibitory concentration (MIC) of thiamphenicol for a wide range of animal and human pathogens as well as genera representative of the human gut flora. The modal MIC_{50} value (minimum inhibitory concentration of thiamphenicol giving complete inhibition of growth of 50% of cultures) was 1.68 μg/ml for 261 bacterial strains isolated from humans. The following species were found to be the most sensitive: *Bacteroides, Fusobacteria, Propionibacteria* and *Actinomyces*. The Committee also noted that 40 μg thiamphenicol/kg food given to mice over 35 days did not alter the intestinal microflora in this species.

The Committee calculated a microbiological ADI for thiamphenicol using the following formula:

$$\text{Upper limit of microbiological ADI} = \frac{MIC_{50} \, (\mu g/g) \times \text{mass of colonic content (g)}}{\text{fraction of oral dose available} \times \text{safety factor} \times \text{human body weight (kg)}}$$

$$= \frac{1.68 \times 220}{0.4 \times 1 \times 60}$$

$$= 15 \, \mu g/kg \text{ bw}$$

In calculating a microbiological ADI the Committee took the following factors into consideration:

- Concentration: $1.68 \, \mu g/ml$ was the modal MIC_{50} for microbiological effects on human intestinal microflora (the density was assumed to be $1 \, g/ml$).

- Availability: the Committee calculated the available portion of thiamphenicol as follows:

  100% ingested   — 60% excreted via urine within 24 hours   = 40% bioavailable in the intestinal tract

  $1 - 0.6 = 0.4$

- Safety factor: the Committee concluded that the data deriving from the microbiological studies (substantial amount of MIC data covering a variety of microorganisms and *in vivo* data from animal studies) provided sufficient information on microbiological effects of thiamphenicol. It therefore adopted a safety factor of 1 in the calculation.

4.   EVALUATION

Taking into account the available toxicological and antimicrobial data and the ADI based on antimicrobial activity, the Committee concluded that the toxicological data provided the most appropriate end-point for the evaluation of thiamphenicol. The Committee established a temporary ADI of 0-6 µg/kg bw for thiamphenicol, based on the NOEL of 1.25 mg/kg bw per day for maternal toxicity in the teratogenicity study in rabbits and a safety factor of 200. The ADI was designated "temporary" because only a summary report of the carcinogenicity study in rats was available. Detailed reports of the carcinogenicity study and the range-finding study used to establish dose levels in that study are required for evaluation in 1999 (see Annex 4).

5.   REFERENCES

Albini, E. (1989). *In vitro* antibacterial activity of thiamphenicol against bacterial strains recently isolated from animal infectious diseases. Unpublished report from Microbiology Laboratory, Zambon Research, Bresso, Milan, Italy. Submitted to WHO by Zambon Group SpA, Bresso, Milan, Italy.

Azzolini, F., Gazzaniga, A., Lodola, E., & Natangelo, R. (1972). Elimination of chloramphenicol and thiamphenicol in subjects with cirrhosis of the liver. *Int. J. Clin. Pharmacol.*, **6**, 130-134.

Bass, R., Oerter, D., Krowke, K., & Spielman H. (1978). Embryonic development and mitochondrial function. III. Inhibition of respiration and ATP generation in rat embryos by thiamphenicol. *Teratology*, **18**, 93-102.

Bichet, N. (1985). Etude de génotoxicité éventuelle sur les hépatocytes de rat par mesure de réparation du DNA. Unpublished report No. 01147F from Sanofi Recherche, Montpellier, France. Submitted to WHO by Zambon Group SpA, Bresso, Milan, Italy.

Bonanomi, L. (1978). Acute toxicity study in rats and mice with thiamphenicol and thiamphenicol glycinate HCl. Unpublished report from Toxicology Department, Zambon Research, Bresso, Milan, Italy. Submitted to WHO by Zambon Group SpA, Bresso, Milan, Italy.

Bonanomi, L. & De Paoli, A.M. (1969). Teratogenicity study of oral thiamphenicol in the rat. Unpublished report from Toxicology Department, Zambon Research, Bresso, Milan, Italy. Submitted to WHO by Zambon Group SpA, Bresso, Milan, Italy.

Bonanomi, L., Pacei, E., & De Paoli, A.M. (1974). Teratogenicity study of oral thiamphenicol in the rabbit. Unpublished report from Toxicology Department, Zambon Research, Bresso, Milan, Italy. Submitted to WHO by Zambon Group SpA, Bresso, Milan, Italy.

Bonanomi, L., De Paoli, A.M., & Gazzaniga, A. (1978). Toxicity study after repeated oral administration of thiamphenicol in the dog. Unpublished report from Toxicology Department, Zambon Research, Bresso, Milan, Italy. Submitted to WHO by Zambon Group SpA, Bresso, Milan, Italy.

Bonanomi, L., Aliverti, V., Ornaghi, F., Losa, M., & Motta, F. (1980). Thiamphenicol perinatal and postnatal toxicity study in the rat. Unpublished report from Toxicology Department, Zambon Research, Bresso, Milan, Italy. Submitted to WHO by Zambon Group SpA, Bresso, Milan, Italy.

Brunaud, M. (1965). Thiamphenicol glycinate hydrochloride. Long-term toxicity in the rabbit by subcutaneous route. Unpublished report from Clin-Byla Laboratories, Paris, France. Submitted to WHO by Zambon Group SpA, Bresso, Milan, Italy.

Cameron, D.M., Crook, D., Brown G., Gopinath, C., Farmer, H., & Offer, J.M. (1990). Z 2041: 4-Week toxicity study in pigs. Unpublished report No. ZBN 12/891518 from Huntingdon Research Centre, Huntingdon, England. Submitted to WHO by Zambon Group SpA, Bresso, Milan, Italy.

Della Bella, D., Bonanomi, L., De Paoli, A.M., & Gazzaniga, A. (1967). Fertility study in male rat treated with thiamphenicol by the oral route. Unpublished report from Toxicology Department, Zambon Research, Bresso, Milan, Italy. Submitted to WHO by Zambon Group SpA, Bresso, Milan, Italy.

Della Bella, D., Ferrari, V., Marca, G., & Bonanomi, L. (1968a). Chloramphenicol metabolism in the phenobarbital induced rat. Comparison with thiamphenicol. *Biochem. Pharmacol.*, **17**, 2381-2390.

Della Bella, D., Bonanomi, L., De Paoli, A.M., & Gazzaniga, A. (1968b). Toxicity study of thiamphenicol after repeated oral administration in the rat. Unpublished report from Toxicology Department, Zambon Research, Bresso, Milan, Italy. Submitted to WHO by Zambon Group SpA, Bresso, Milan, Italy.

Ferrari, V. & Della Bella, D. (1974). Comparison of chloramphenicol and thiamphenicol metabolism. *Postgrad. Med. J.*, **50**, 17-22.

Gazzaniga, A. (1974). Thiamphenicol - pharmacokinetics and metabolism in animals. Unpublished report from Toxicology Department, Zambon

Research, Bresso, Milan, Italy. Submitted to WHO by Zambon Group SpA, Bresso, Milan, Italy

Kelly, C.M. & Daly I.W. (1990). A four week oral toxicity study in the dog via capsule administration with Z 2041. Unpublished report No. 88-3391 from Bio/dynamics Inc., East Millstone, NJ, USA. Submitted to WHO by Zambon Group SpA, Bresso, Milan, Italy.

Kelly, C.M. & Daly, I.W. (1991). A six month oral toxicity study in the dog via capsule administration with Z 2041 with a two month recovery period. Unpublished report No. 88-3392 from Bio/dynamics Inc., East Millstone, NJ, USA. Submitted to WHO by Zambon Group SpA, Bresso, Milan, Italy.

Kelly, J.P. & Kaufman, D.W. (1989). Anti-infective drug use in relation to the risk of agranulocytosis and aplastic anemia: A report from the international agranulocytosis and aplastic anemia study. *Arch. Intern. Med.,* **149**, 1036-1040.

Laplassote, J. (1962). Recherches expérimentales sur le thiopénicol: activité antibactérienne, concentrations humorales, élimination. Comparaison avec le chloramphénicol. *Therapie,* **16**, 104-108.

Marca, G. & Bonanomi, L. (1979). Gene conversion and mitotic crossing-over in *Saccharomyces cerevisiae* 6117 with thiamphenicol with and without metabolic activation. Unpublished report from Toxicology Department, Zambon Research, Bresso, Milan, Italy. Submitted to WHO by Zambon Group SpA, Bresso, Milan, Italy.

Maekawa, A. (1996). Summary report on two-year rat carcinogenicity study of thiamphenicol. Submitted by Sasaki Institute, Tokyo, Japan to the Japanese Ministry of Health and Welfare.

Marubini, M., Motta, F., Cerioli, A., & Finn, J.P. (1991). Z 2041. Thirteen week oral toxicity study in rats followed by a recovery period of 8 weeks. Unpublished report No. 1219 from Toxicology Department, Zambon Research, Bresso, Milan, Italy. Submitted to WHO by Zambon Group SpA, Bresso, Milan, Italy.

Mosesso, P. & Driedger, A. (1989). Chromosome abberations in human lymphocytes cultured *in vivo*. Unpublished report No. LSC-RTC 116028-M-02989 from Life Science Research Roma Toxicology Centre SpA, Rome, Italy. Submitted to WHO by Zambon Group SpA, Bresso, Milan, Italy.

Murray, T.R., Downey, K.M., & Yunis, A.A. (1983). Chloramphenicol-mediated DNA damage and its possible role in the inhibitory effects of chloramphenicol on DNA synthesis. *J. Lab. Clin. Med.,* **102**, 926-932.

Nijhof, W. & Kroon, A.M. (1974). The interference of chloramphenicol and thiamphenicol with the biogenesis of mitochondria in animal tissues. A possible clue to the toxic action. *Postgr. Med. J.*, **50**(Suppl. 5), 53-59.

Pinasi, C., Ferrini, S., & Ceriani, D. (1990a). Mutagenicity evaluation of Z 2041 in the Ames Salmonella-microsome plate test. Unpublished report No. 1231 from Toxicology Department, Zambon Research, Bresso, Milan, Italy. Submitted to WHO by Zambon Group SpA, Bresso, Milan, Italy.

Pinasi, C., Ferrini, S., & Ceriani, D. (1990b). Mutagenicity evaluation of Z 2041 in Chinese hamster V79 test. Unpublished report No. 1243 from Toxicology Department, Zambon Research, Bresso, Milan, Italy. Submitted to WHO by Zambon Group SpA, Bresso, Milan, Italy.

Pinasi, C., Ferrini, S., & Ceriani, D. (1990c). Genotoxicity evaluation of Z 2041 in the "*in vivo*" mouse micronucleus assay. Unpublished report No. 1233 from Toxicology Department, Zambon Research, Bresso, Milan, Italy. Submitted to WHO by Zambon Group SpA, Bresso, Milan, Italy.

Poli, G. (1994). Qualitative and quantitative changes in the mouse intestinal microflora induced by chronic thiamphenicol at maximum residual level (MRL) dose. Unpublished report from Facolta di Medicina Vetarinaria, Universita di Milano, Milan, Italy. Submitted to WHO by Zambon Group SpA, Bresso, Milan, Italy.

Redgrave, V.A., Cameron, D.M., Anderson, A., & Maxwell, J.G. (1991). Blood concentrations of thiamphenicol in pigs following dietary administration of Z 2041. Unpublished report No. ZBN 11/901351 from Huntingdon Research Centre, Huntingdon, England. Submitted to WHO by Zambon Group SpA, Bresso, Milan, Italy.

Roberts, N.L., Cameron, D.M., Redgrave, V.A., Crook, D., & Brown G. (1989). Z 2041. Tolerance in pigs. Unpublished report No. ZBN 10/89188 from Huntingdon Research Centre, Huntingdon, England. Submitted to WHO by Zambon Group SpA, Bresso, Milan, Italy.

Schioppacassi, G. (1992). Activity of thiamphenicol on indigenous microogranisms of gastrointestinal tract. Unpublished report No. DBI 1/92 from Zambon Research, Bresso, Milan, Italy. Submitted to WHO by Zambon Group SpA, Bresso, Milan, Italy.

Simpson, L.O., Aarons, I., & Howie, J.B. (1979). Thiamphenicol and lupus nephritis. The effects of long-term therapy on kidney function and pathology: a pilot study. *Br. J. Exp. Pathol.*, **60**, 45-57.

Sisti, R. (1994). Z 2041. Oral embryotoxicity study in rabbits. Unpublished report No. RTC 4195/T/128/94 from Research Toxicology Centre SpA, Rome, Italy. Submitted to WHO by Zambon Group SpA, Bresso, Milan, Italy.

Sutter, V.L. & Finegold, S.M. (1976). Susceptibility of anaerobic bacteria to 23 antimicrobial agents. *Antimicrob. Agents Chemoter.*, **10**, 736-752.

TAP Pharmaceuticals Inc. (1987). Issues relating to the safety of thiamphenicol with special attention to the matter of aplastic anemia. Unpublished report from TAP Pharmaceuticals Inc., Corte Mader, Canada. Submitted to WHO by Zambon Group SpA, Bresso, Milan, Italy.

Tomoeda, M. & Yamamoto, K. (1981). The hematologic adverse reaction experience with thiamphenicol in Japan. In: Safety Problems Related to Chloramphenicol and Thiamphenicol Therapy, Raven Press, New York, pp. 103-110.

Uesugi, T., Ikeda, M., Hori, R., Katayama, K., & Arita, T. (1974). Metabolism of thiamphenicol and comparative studies of its urinary and biliary excretion with chloramphenicol in various species. *Chem. Pharm. Bull.*, **22**, 2714-2722.

Yunis, A.A. & Gross, M.A. (1975). Drug-induced inhibition of myeloid colony growth: protective effect of colony stimulating factor. *J. Lab. Clin. Med.*, **86**, 499-504.

TILMICOSIN

First draft prepared by
Dr G. Roberts
Commonwealth Department of Health and Family Services
Canberra, Australia

1.   EXPLANATION

Tilmicosin is a macrolide antibiotic with the chemical name of 20-deoxo-20-(3,5-dimethyl piperidin-1-yl) desmycosin. It is structurally similar to tylosin. Tilmicosin is a mixture of one cis and two trans isomers in the approximate ratio 85:15.

Tilmicosin is available as an injectable formulation for the treatment of respiratory diseases in cattle and sheep (10 mg/kg bw) and as a feed premix for the treatment and control of respiratory diseases in pigs (200 to 400 mg/kg in the feed). It had not been previously evaluated by the Committee. The molecular structure of tilmicosin is shown below.

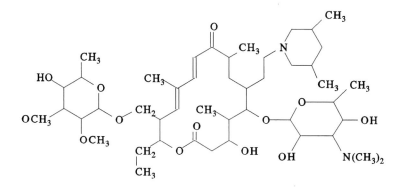

2. BIOLOGICAL DATA

2.1 Biochemical aspects

2.1.1 Excretion

Pigs were administered a dose of 110 mg ^{14}C-tilmicosin in the diet over the course of one day. The recovery of radioactivity was 15% in the urine and 80% in the faeces (Giera & Thomson, 1986).

In two studies, pigs were given a dose of 154 or 400 mg ^{14}C-tilmicosin in the diet following a similar dose given for 5 days. The recovery of radioactivity was 4 to 6% in urine and 62 to 75% in faeces. Radioactivity was detected in the bile but was not quantified (Donoho & Thomson, 1988; Donoho et al., 1993).

2.1.2 Biotransformation

2.1.2.1 Rats

Tilmicosin, labelled with ^{14}C in both the desmycosin macrolide ring and the piperidine ring, was given orally to 15 male and 15 female Fischer-344 rats. The dosage was 20 mg/kg bw per day for 3 days. In the liver, radiolabel corresponded to tilmicosin and a desmethyl derivative, T1(demethylated in the mycaminose ring). The single radioactive substance identified in urine was unchanged tilmicosin, while in faeces the major peak was parent compound with lesser amounts of desmethyl tilmicosin and a high molecular weight compound known to be present as an impurity in the administered substance, T2 (consisting of two macrolide rings and one piperidine ring) (Donoho, 1988).

Fischer-344 rats (10 males and 10 females) were given gavage doses of 50 mg/kg bw per day ^{14}C-tilmicosin for 5 days. An analysis of faecal radioactivity for the presence of the sulfate metabolite that was found in the faeces of pigs revealed the presence of a similar compound, but quantification was not undertaken (Donoho & Kennington, 1993).

2.1.2.2   Sheep

Beulah cross lambs were administered a single subcutaneous dose of 20 mg/kg bw per day ^{14}C-tilmicosin. The major radioactive component in the liver, kidneys and urine was the parent drug, together with lesser amounts of T1 and T2, and minor amounts of other unidentified substances (Hawkins *et al.*, 1993).

2.1.2.3   Pigs

Pigs were fed diets containing ^{14}C-tilmicosin, which resulted in daily doses of 400 mg, for 5 days. Of the radiolabel in liver and kidney, approximately 60 to 70% was in the form of the parent drug and there were small amounts of T1. Similarly in urine and faeces, there were high levels of tilmicosin and low levels of T1. A further metabolite was detected that comprised 14% of faecal and 25% of urinary radioactivity and that was consistent with reduction of one double bond in the macrolide ring followed by sulfation (Donoho *et al.*, 1993).

2.1.2.4   Cattle

In a summary of results obtained in cattle injected with ^{14}C-tilmicosin, it was reported that the radioactivity profile in the liver of treated rats was similar to that in the faeces. In animals treated with a highly purified sample of tilmicosin, metabolite T2 was not detected in the liver, suggesting that its presence was a result of direct administration as a component of the drug substance. Radioactivity in the kidneys was essentially in the form of unchanged tilmicosin (Donoho, 1988).

2.2   Toxicological studies

2.2.1   Acute studies

Major signs of toxicity in mice and rats were leg weakness, hypoactivity, lethargy, ataxia, poor grooming and coma. Monkeys given 20 mg/kg vomited during the first day but were subsequently normal. The single monkey given 30 mg/kg vomited, exhibited hypoactivity, laboured respiration, vocalization and ataxia, and died within 2 hours.

Table 1. Results of acute toxicity studies with tilmicosin

Species (strain)	Route	Vehicle	Sex	LD$_{50}$[3] (mg/kg bw)	Reference
Mouse (ICR)	sc	aqueous	M F	97 109	Jordan *et al.*, 1986a
Rat[1] (Sprague-Dawley)	oral	aqueous	M F	850 800	Jordan *et al.*, 1986b
Rat[2] (Fischer-344)	oral	aqueous	M & F	> 2000	Piroozi *et al.*, 1993
Rat (Fischer-344)	sc	aqueous	M F	185 440	Jordan *et al.*, 1986c
Rat (Fischer-344)	inhalation	aerosol	M & F	> 494 (0/20) < 4053 (14/20)	Jordan *et al.*, 1987
Rabbit (NZ White)	dermal	-	M & F	> 5000	Jordan *et al.*, 1987
Monkey (Rhesus)	im	aqueous	?	> 20 (0/2) 30 (1/1)	Jordan & Shoufler, 1990
Sheep	sc iv	aqueous aqueous	? ?	> 150 (0/5) > 7.5 (1/5)	Cochrane & Thomson, 1990

[1] Animals were fasted
[2] Animals were non-fasted
[3] Figures in brackets represent the incidence of deaths

New Zealand white rabbits (five males and five females) received 5000 mg/kg bw tilmicosin onto clipped skin for a period of 24 hours, under a non-occlusive dressing. There were no deaths and no signs of skin irritation (Jordan *et al.*, 1987).

A volume of 0.1 ml (17 mg) of tilmicosin was instilled into one eye of three male and three female New Zealand White rabbits. Conjunctival hyperaemia and chemosis were noted for several days, clearing by the end of one week (Jordan *et al.*, 1987).

Hartley albino guinea-pigs (6 to 12 females) were given 10 intracutaneous induction doses of tilmicosin (0.05 to 0.1 ml of a 50 mg/ml solution). Fourteen days after the induction phase each animal was injected intracutaneously with 0.05 ml tilmicosin. This challenge dose did not result in an enhancement of skin reactions, suggesting that a skin sensitization response had not been elicited (Jordan *et al.*, 1989b).

2.2.2   Short-term toxicity studies

2.2.2.1   Rats

Groups of 20 male and 20 female Fischer-344 rats were given gavage doses of 0, 50, 175 or 600 mg/kg bw per day tilmicosin (purity 88%) in an aqueous vehicle. Drug administration was continued for 2 weeks, during which there was no mortality or clinical signs of toxicity. Food consumption was lower in males and females at 600 mg/kg and body weight gain was depressed in males throughout the study and in females during the first few days only.

In the 600 mg/kg bw per day groups, there were increases in haematocrit owing to an increase in corpuscular volume and in serum levels of alanine aminotransferase and hepatic *p*-nitroanisole *O*-demethylase activity. Thrombocyte counts were slightly decreased at this dose. Urinalysis was unaffected. At the highest dose, there were slight decreases in the weights of kidneys, spleen and ovaries, increased adrenal weight and multifocal inflammation in the liver of some animals. All treated animals exhibited caecal distension, which is a typical response to high doses of an antibiotic compound (Jordan, 1986).

Groups of 20 male and 20 female Sprague Dawley (Crl:CD) rats received 0, 50, 250 or 1000 mg/kg bw per day tilmicosin (purity 88%) in an aqueous vehicle, by gavage for 3 months. Signs of overt toxicity were noted at 1000 mg/kg and included thinness, ventral soiling, chromorhinorrhea, chromodacryorrhea, alopecia and poor grooming. Mortality was increased at this highest dose and in females given 250 mg/kg bw per day.

Although food consumption was depressed in the 1000 mg/kg bw per day group (males only), body weight gain was lower in males and females at 1000 mg/kg bw per day and females at 250 mg/kg bw per day. At 1000 mg/kg bw per day, serum alanine aminotransferase was increased in males, blood urea nitrogen (BUN) was increased in males and females, urinary pH was slightly lower in females and the presence of occult blood was greater in males and females. There were no effects on ophthalmology, haematology or hepatic *p*-nitroanisole *O*-demethylase activity.

The weights of kidneys, liver and heart were increased in females at 1000 mg/kg bw per day and adrenal weight was increased at 250 mg/kg bw per day (females only) and 1000 mg/kg bw per day (males and females). Gross pathology revealed caecal enlargement or distension at 250 mg/kg bw per day or more and small spleens in a few rats at 1000 mg/kg bw per day. Slight nephrosis was noted in two males in each of the 250 and 1000 mg/kg bw per day groups. Other pathological alterations were seen only in males and females of the 1000 mg/kg bw per day group, i.e. hypertrophy of the zona fasciculata of the adrenal cortex in most rats, necrosis of individual skeletal muscle fibres in some rats; myocardial degeneration was increased and lymphoid depletion was evident in the spleen and thymus of some animals. The NOEL was 50 mg/kg bw per day (Jordan, 1988).

2.2.2.2   Dogs

Beagle dogs (two males and two females per group) were exposed by inhalation to aerosols of 0, 12, 47 or 251 mg/m^3 tilmicosin (purity 83%). The treatment was administered for 4 hours per day on 12 days out of a 16-day period (excluding weekends). Particle median equivalent aerodynamic diameters were between 1.2 and 1.5 μm. The mean serum concentrations of tilmicosin were 0.23 to 0.33 and 1.67 to 2.46 μg/ml at 47 and 251 mg/m^3 respectively. Drug levels at 12 mg/m^3 were below the detection limit of 0.1 μg/ml.

One male given 251 mg/m^3 died on the final day of exposure but no other overt signs of toxicity were observed. Heart rates were increased in dogs at 251 mg/m^3. Food intake, body weight, haematology and blood chemistry were unaffected. At autopsy, lung weight was increased in females given 251 mg/m^3 and inflammation was noted in the respiratory tract at 47 mg/m^3 (females only) and 251 mg/m^3 (males and females) (Jordan et al., 1991).

Groups of four male and four female beagle dogs were given 0, 6, 20 or 70 mg/kg bw per day tilmicosin (purity 88%) in capsules for 3 months. The daily dose was administered in two equal amounts, 6 hours apart. Serum concentrations of tilmicosin were measured 3 hours and 24 hours after the first of the two daily doses in weeks 2, 3, 5, 9 and 13. At increasing doses, the mean concentrations were 0.1 to 0.2, 0.74 to 1.49 and 3.24 to 6.05 μg/ml at 3 hours and 0, 0.26 to 0.59 and 1.72 to 3.96 μg/ml at 24 hours.

Half the 70 mg/kg bw per day males and females died during the first month. Clinical signs prior to death were pale mucous membranes in two dogs and ataxia in one dog. Food intake, body weight, haematology and urinalysis were unaffected. Serum alanine aminotransferase activity was progressively increased at 70 mg/kg bw per day, while hepatic p-nitroanisole O-demethylase activity was increased in females at this level.

Two of the four surviving dogs at 70 mg/kg bw per day showed bilateral multifocal areas of subretinal fluid concentrated along arterioles in the tapetal region of the eye. The changes were claimed to be consistent with those associated with systemic hypertension. One of these dogs also had retinal degeneration and the other showed miosis with normal pupillary light responses.

Heart rate increases were dose related, being marked at 70 mg/kg bw per day and moderate to severe at 20 mg/kg bw per day. The increase in males given 6 mg/kg bw per day was not significant. At the highest dose, examination of the electrocardiogram revealed depression of the ST segment.

At 70 mg/kg bw per day, heart weight was increased in males and females, and liver and kidney weights were increased in females. Gross necropsy showed small spleens in the females that died during the study and an enlarged heart in a surviving male at 70 mg/kg bw per day. Slight diffuse mucosal oedema was seen in the gall bladders of two dogs at 70 mg/kg bw per day. Despite the physiological changes in the eyes and heart, there were no associated pathological alterations. The NOEL was 6 mg/kg bw per day (Jordan, 1987).

Groups of four male and four female beagle dogs were given 0, 4, 12 or 36 mg/kg bw per day tilmicosin (purity 86%) in capsules for 12 months. The daily dose was administered in two equal amounts, 6 hours apart. There were no deaths. Peripheral redness was seen sporadically in some animals of all groups, in particular at 12 and 36 mg/kg bw per day.

Body weight gains were lower at ≥ 12 mg/kg bw per day. There were no effects on ophthalmology, haematology, blood chemistry and urinalysis. Heart rates were markedly increased at 36 mg/kg bw per day with sporadic depression of the ST segment in some dogs.

Heart weight was increased at 36 mg/kg bw per day and four males and one female at this dose showed enlarged hearts. Mild chronic dermatitis was noted in the external ears of dogs from all treated groups. The findings included minimal thickening of the epidermis, foci of acantholysis and inflammatory cell infiltration in the dermis. These changes were only slight and non-dose-related, and therefore of questionable relationship to treatment. The NOEL was 4 mg/kg bw per day (Jordan & Bernhard, 1989).

2.2.3   Long-term toxicity/carcinogenicity studies

Tilmicosin has not been tested in toxicological studies longer than 12 months in duration and hence the carcinogenic potential of this drug has not been directly determined. There are a number of observations which are relevant in assessing the carcinogenicity of tilmicosin:

a)   The results of available toxicological studies with tilmicosin have not identified lesions or proliferative changes which could be considered suggestive of potential neoplasia.

b)   Tilmicosin has been tested in a wide range of genotoxicity assays. All results were negative and it was concluded that the compound has no genotoxic activity.

c)   Tilmicosin is a macrolide antibiotic. This class of chemical has been in widespread usage in humans for many years but there is no evidence of carcinogenicity. Tylosin is the closest structural analogue and this chemical was reviewed by the thirty-eighth meeting of JECFA (Annex 1, reference 97).

2.2.3.1 Reconsideration of tylosin tumorigenicity

Tylosin was tested in 2-year feeding studies in rats. The findings revealed an association between drug administration at doses of 150 and 300 mg/kg bw per day in the diet and an increased incidence of pituitary adenomas in males. While it was claimed that the increase in the incidence of tumours was an indirect result of the ability of tylosin to increase survival and weight gain, supporting data were not available to the Committee and an ADI was not established.

New information, not considered at the thirty-eighth meeting, indicates that spontaneous pituitary tumours occur in male rats at variable rates with increased rates occurring in animals with higher body weights (Gries & Young, 1982). In the tylosin studies, 12-month body weights were somewhat higher in treated males rats and mortality was increased in control animals due to respiratory infection. The highest observed incidence of tumours in tylosin-treated groups was comparable to the upper limit of the historical control range (23%) in the test facility.

2.2.4 Reproductive toxicity studies

2.2.4.1 Rats

A dose-ranging study was carried out in groups of 10 female Sprague-Dawley (Crl:COBS CD) rats given gavage doses of 0, 50, 125, 250, 500 or 750 mg/kg bw per day tilmicosin (purity unknown). The females were treated from 14 days before mating with untreated males until the end of study on post-partum day 4.

Excess salivation was dose-related at $\geq$ 250 mg/kg bw per day. Chromodacryorrhea, urine-stained fur and alopecia were seen in some rats at 750 mg/kg bw per day; the three most severely affected died during the study. Prior to death, these animals exhibited reduced food intake, lost weight and were emaciated. Body weight gain was depressed in surviving rats at 750 mg/kg bw per day during the pre-mating period while food intake was reduced at $\geq$ 250 mg/kg bw per day in week 1 and at 750 mg/kg bw per day in week 2. The pregnancy rate was decreased at $\geq$ 500 mg/kg bw per day. Duration of gestation, litter size and weight, pup survival and weight gain to post-partum day 4 were unaffected (Dearlove *et al.*, 1987).

Groups of 30 male and 30 female Sprague Dawley (Crl: COBS CD BR) rats were administered 0, 10, 45 or 200 mg/kg bw per day tilmicosin (purity 87%) by gavage in an aqueous vehicle. Treatment commenced 70 days (males) and 14 days (females) before the first mating period and was continued through two litters per generation for two generations. F_1

litters were culled to five pups per sex on post-partum day 4 and F_{1b} pups (40 males and 40 females per group) were bred to produce the following generation. F_{2a} and F_{2b} pups were killed on post-partum day 4.

In adult animals, salivation was noted in males and females at 200 mg/kg bw per day, but there were no other signs of toxicity. Body weight gain and food consumption were depressed in females of the 45 and 200 mg/kg bw per day groups during the pre-mating period only. During the production of the four litters in this study there were no effects on mating performance, pregnancy rates, duration of gestation, litter size and weight and the weight gain of offspring. In both F_1 litters at 200 mg/kg bw per day, pup mortality was slightly increased up to post-partum day 4, but the finding was not duplicated in either F_2 litter. The NOEL was 10 mg/kg bw per day in adult rats (Christian & Hoberman, 1989).

2.2.5 Special studies on embryotoxicity and teratogenicity

2.2.5.1 Rats

Groups of 25 presumed pregnant female Sprague Dawley (Crl: CD) rats were administered gavage doses of 0, 10, 70 or 500 mg/kg bw per day tilmicosin (purity 86%) in an aqueous vehicle. The dams were treated on gestation days 6 to 15 and were killed on gestation day 20. Increased salivation was seen at 70 and 500 mg/kg bw per day and alopecia at 500 mg/kg bw per day. There were no deaths or abortions. Body weight gain was reduced at 70 and 500 mg/kg bw per day, and food intake was reduced at 500 mg/kg bw per day, during gestation days 6 to 10.

The number of resorption or live fetuses, fetal weight, sex ratio and the incidence of fetal malformations were similar between groups. The incidences of total skeletal and visceral anomalies were increased in treated groups, but there was no dose-response relationship and the findings were within the historical control values for the laboratory. The NOEL for maternotoxicity was 10 mg/kg bw per day (Jordan & Higdon, 1988).

2.2.5.2 Rabbits

Groups of 15 presumed pregnant female Japanese White-NIBS rabbits were given gavage doses of 0, 8, 19 or 48 mg/kg bw per day tilmicosin (purity unknown) in an aqueous vehicle. Does were treated on gestation days 6 to 18 and were killed on gestation day 28. One female given 48 mg/kg bw per day aborted on gestation day 26 and died. Reduced faeces were seen at 19 and 48 mg/kg bw per day with only a transient effect at 8 mg/kg bw per day. During the treatment period food intake was depressed in a dose-related manner and body weight loss was observed at 19 and 48 mg/kg bw per day.

Fetal and placental weight tended to decrease at 19 and 48 mg/kg bw per day but did not achieve statistical significance. There were no meaningful effects on the incidence of resorptions or fetal deaths or on sex ratio. Open eyelids were observed in 11/91 and 16/68 fetuses from the 19 and 48 mg/kg bw per day groups, respectively, and some of these fetuses showed cleft palate or club foot. The affected fetuses had low body weights and the seven litters involved were derived from does that had lost body weight during drug administration. Skeletal examination revealed retardation of fetal development at 19 and 48 mg/kg bw per day. Similar effects have been seen in this laboratory in dietary restricted rabbits, suggesting the effects on the fetus were secondary to maternal malnutrition (Noda, 1993).

2.2.6   Special studies on genotoxicity

Table 2. Results of genotoxicity studies on tilmicosin

Test system[1]	Test object	Concentration	Results	Reference
Reverse mutation[2]	S. typhimurium TA98, TA100, TA1535, TA1537, TA1538	1-100 µg/plate (± S9)	negative	Jordan et al., 1986f
Reverse mutation[2]	E. coli WP2uvrA⁻	1-100 µg/plate (± S9)	negative	Garriott et al., 1993
Forward mutation[2]	L5178Y mouse lymphoma cells	100-900 µg/ml (- S9) 200-1000 µg/ml (+ S9)	negative	Jordan et al., 1986e
Forward mutation[2]	HGPRT⁺ Chinese Hamster ovary cells	25-250 µg/ml (- S9) 50-300 µg/ml (+ S9)	negative	Jordan et al., 1989c
Unscheduled DNA synthesis assay	primary cultures of rat hepatocytes	0.5-10 µg/ml	negative	Jordan et al., 1985
Sister chromatid exchange assay	Chinese hamster bone marrow	200-1800 x 1 mg/kg bw oral	negative	Jordan et al., 1986d, 1989a
Chromosome aberrations	rat bone marrow	180-1800 x 1 and 100-1000 x 5, mg/kg bw per day oral	negative	Jordan & Ivett, 1989

[1]   Positive controls used
[2]   Both with and without liver microsomal activation

2.2.7 Special study on the immune system

Groups of eight male CD-1 mice were given gavage doses of 0, 10, 250, 500 or 1000 mg/kg bw per day tilmicosin (purity 87%) in an aqueous vehicle. Drug treatment was continued for 10 days. After the third dose mice were injected intravenously with sheep red blood cells. At the end of the study, serum from each mouse was assayed for antibodies (haemagglutinin). There was one death at 500 mg/kg bw per day and six at 1000 mg/kg bw per day, but antibody production was not affected at any level (McGrath *et al.*, 1988a).

2.2.8 Special studies on microbiological activity

2.2.8.1 *In vitro*

The antibacterial activity of tilmicosin was determined against a range of organisms representative of human intestinal flora. Cultures were initiated using inoculum sizes of between 3×10^5 and 1.6×10^6 cells per spot on blood agar plates and incubated anaerobically at 37 °C.

Microorganisms (number of strains)	MIC (μg/ml)
Standard NCTC strains:	
Bifidobacterium (n=5)	0.004-0.015
Peptostreptococcus (n=3)	0.0075-0.5
Bifidobacterium isolated from healthy babies (n=6)	0.0075-> 64
Peptostreptococcus isolated from sites of clinical disease (n=6)	0.015-4

The median MIC values for the clinical isolates of *Bifidobacterium* and *Peptostreptococcus*, respectively, were 0.015 and 0.5 μg/ml. The above results were obtained at pH 7.7. At pH 6.6 the MIC values were one to two orders of magnitude higher (Scott *et al.*, 1993).

The antibacterial activity of tilmicosin was examined against a number of organisms used in industrial food processing and originally obtained from dairy products. Cultures were initiated in blood agar plates using three strains of *Propionibacterium* and four strains of *Lactobacillus*.

Under anaerobic conditions the MICs were between 8 and 64 μg/ml and under microaerophilic conditions the values were between 4 and 32 μg/ml (McLaren, 1994).

2.2.8.2   Rats

In two separate studies, female germ-free rats were orally dosed with 1 ml of 10% (w/v) pooled human faecal suspension. Three weeks later groups of two males and two females were given gavage doses of 0, 0.12 or 0.4 mg/kg bw per day tilmicosin (purity 87%) for 5 days. Rat faecal samples were collected before treatment, daily during treatment and once during each of the 2 weeks following cessation of dosing. Body weight gain remained unaffected.

The animals in each study received faecal suspensions obtained from different donors. Results from different studies varied widely, possibly reflecting differences between the bacterial populations used. Total anaerobes were not reduced by tilmicosin treatment. Total enterobacterial counts and the proportion of enterobacteria with respect to total anaerobe counts were transiently increased at 0.4 mg/kg bw per day in one study only. Tilmicosin-resistant enterobacteria were not significantly increased in either study. Spiramycin, at a dose of 0.5 mg/kg bw per day, increased total enterobacterial counts and the number of spiramycin-resistant enterobacteria; the latter appeared to remain high after the cessation of treatment. The NOEL for tilmicosin was 0.4 mg/kg bw per day (Rumney, 1993).

2.2.9   Special studies on pharmacology

Table 3. Results of pharmacological assays with tilmicosin

Test system	Doses	Results	Reference
Isolated guinea-pig ileum	0.0009 to 90 μg/ml	No effect on non-stimulated organ. 90 μg/ml slightly inhibited acetylcholine- and angiotensin-induced contraction and significantly inhibited electrically stimulated contraction	Williams *et al.*, 1988
Isolated rat uterus (estrogen primed)	0.0009 to 90 μg/ml	No effect on non-stimulated tissue and no change to oxytocin-induced contractions	Williams *et al.*, 1988
Isolated rat vas deferens	0.0009 to 90 μg/ml	No effect on non-stimulated tissue and no change to noradrenaline-induced contractions	Williams *et al.*, 1988

Table 3 (contd).

Test system	Doses	Results	Reference
Isolated guinea-pig atria	0.0009 to 900 μg/ml	In spontaneously beating tissue, ≥ 90 μg/ml decreased force of contraction; and 900 μg/ml increased rate of contraction. 90 μg/ml inhibited isoprenaline-, noradrenaline- and Bay K8644 (calcium agonist)-induced force and rate of contraction, and histamine-induced rate of contraction	Williams et al., 1988; Jordan et al., 1990
Conscious dogs	0.25, 1, 2.5 or 5 mg/kg iv	Heart rate increased at ≥ 1 mg/kg. Left ventricular function decreased at ≥ 0.25 mg/kg. Aortic pulse pressure decreased at ≥ 1 mg/kg. ECG revealed "Ventricular Alternans" (ventricular contraction fails to achieve adequate pressure to open aortic valve and cause an arterial pulse).	Jordan & Sarazan, 1991; Sarazan et al., 1993
Anaesthetised dogs	0.5, 1 or 5 mg/kg iv	Heart rate increased, stroke volume, stroke work index and cardiac output decreased at ≥ 1 mg/kg. At 5 mg/kg 2/3 dogs developed second degree heart block	Jordan et al., 1988
CD-1 mice	10, 100, 500 or 1000 mg/kg oral	Piloerection and grasping loss at ≥ 500 mg/kg. Ptosis at 1000 mg/kg. No effect on electroshock- or pentylenetetrazol-induced seizures or on body temperature. Hexobarbital-induced sleep time prolonged by 500 (x1) and 1000 (x3) mg/kg	Jordan et al., 1989d
Sprague-Dawley rats	10, 30, 90 or 270 mg/kg oral	Urine volume and creatinine reduced at 270 mg/kg with increases in osmolality and electrolytes.	McGrath et al., 1988b

2.3   Observations in humans

A total of 241 human exposures to tilmicosin (Micotil) were reported to the Rocky Mountain Poison Center from May 1992 to May 1993. Needlesticks and scrapes (n=112) and accidental injection (n=43) caused either no effect or local reactions including soreness, numbness, stinging, swelling, redness, burning and stiffness. Some injected subjects experienced anxiety, sweating, headache and lightheadedness. Skin exposures (n=50) resulted in redness and tingling of the skin, and eye exposures (n=11) resulted in stinging and swelling. Persons accidentally ingesting the drug (n=39) reported either no symptoms or bitter taste, nausea, numbness of lips and tongue, vomiting, thirst and headache (Montanio & Dart, 1993).

Over a 30-month period, 36 cases of accidental exposure to tilmicosin (Micotil) were reported to the Ontario Regional Poison Information Centre. Percutaneous injection (n=26) always resulted in pain at the site and seven subjects mentioned a variety of local reactions consistent with an irritant action. One person showed "peaked T waves" 30 minutes following the injection of 1 ml into his arm, but ECG changes were not noted in other subjects. The remaining subjects received splashes into the mouth or eyes or onto the skin. Unpleasant taste, a burning sensation on the hard palate and ocular irritation were each recorded in one person (McGuigan, 1994).

3.   COMMENTS

The Committee considered toxicological data on tilmicosin, including the results of studies on acute and short-term toxicity, pharmacokinetics, metabolism, reproductive toxicity, teratogenicity, genotoxicity, antimicrobial activity and pharmacology. The Committee also considered observations in humans accidentally exposed to tilmicosin.

Tilmicosin is absorbed from the gastrointestinal tract as shown by the recovery of radiolabelled compound in urine and/or bile of orally dosed pigs (110 to 400 mg per animal), and the presence of dose-related serum concentrations of tilmicosin dosed orally in dogs for 3 months. However, the results from available studies did not allow any conclusion to be reached concerning the extent of absorption. The metabolites detected in a range of farm animals were also found in rats, which suggests that the rat is a suitable model for determining the potential toxicological hazards posed by tilmicosin.

The LD_{50} in fasted rats was 800 to 850 mg/kg bw, but toxicity was substantially lower in non-fasted animals, among which there were no deaths following administration of a single dose of 2000 mg tilmicosin/kg bw. Studies in a range of pharmacological models identified depression of cardiac muscle contractility and reduced heart function as the main adverse effects.

Rats were administered oral doses of 50, 250 or 1000 mg/kg bw per day for 3 months. At 1000 mg/kg bw per day, animals exhibited chromorhinorrhea, chromodacryorrhea, alopecia, poor grooming, reduced food consumption and they appeared thin. Body-weight gain was reduced and mortality was increased in females at 250 mg/kg bw per day and in both sexes at 1000 mg/kg bw per day. Toxic effects on the kidney and liver were indicated by increased organ weights, increased serum levels of alanine aminotransferase and blood urea nitrogen, and the presence of blood in the urine. However, pathological examination of these organs showed only slight nephrosis in two males in each of the groups given 250 or 1000 mg/kg bw per day. At the highest dose, hypertrophy of the zona fasciculata of the adrenal cortex was noted in most rats, and myocardial degeneration, necrosis of individual skeletal muscle fibres and lymphoid depletion in the spleen and thymus were seen in some rats. The NOEL was 50 mg/kg bw per day.

Dogs were given oral doses of 6, 20 or 70 mg/kg bw per day for three months. Half of the animals given 70 mg/kg bw per day died during the first month. Heart rate was increased at 20 and 70 mg/kg bw per day and at the higher dose there was depression of the ST segment of the electrocardiogram. Ophthalmoscopic examination of the eyes showed bilateral areas of subretinal fluid concentrated along arterioles in the tapetal region of two of the four surviving dogs at 70 mg/kg bw per day. The NOEL was 6 mg/kg bw per day.

Dogs were given oral doses of 4, 12 or 36 mg/kg bw per day for twelve months. Body weight gain was depressed at 12 and 36 mg/kg bw per day. Heart rate was markedly increased at 36 mg/kg bw per day, with occasional depression of the ST segment of the electrocardiogram. Hearts were enlarged at 36 mg/kg bw per day, but there were no associated pathological changes. The NOEL was 4 mg/kg bw per day.

In a two-generation reproductive toxicity study, rats were administered oral doses of 10, 45 or 200 mg tilmicosin/kg bw per day. Body weight gain and food consumption were depressed in females given 45 or 200 mg/kg bw per day, but only during the pre-mating period. Fertility and reproduction parameters were unaffected at any dose level. At 200 mg/kg bw per day, mortality in pups was slightly increased in both first-generation litters up to post-partum day 4, but not in either second-generation litter. The NOEL was 10 mg/kg bw per day.

Developmental toxicity studies were reported in rats and rabbits. Rats were dosed with 10, 70 or 500 mg tilmicosin/kg bw per day by gavage. Growth and development in rat fetuses were not affected by treatment, but increased salivation and reduced body weight gain of dams were observed at 70 and 500 mg/kg bw per day. The NOEL for maternal toxicity in rats was 10 mg/kg bw per day. In the study in rabbits, there were dose-related signs of toxicity in the does at all doses (8, 19 and 48 mg/kg bw per day). Severely affected members of the groups given 19 or 48 mg/kg bw per day produced offspring that had open eyelids and low incidences of cleft palate or club foot. These fetuses had low body weights and were clearly retarded in development. The effects were attributed to treatment-related malnutrition in maternal animals, which is commonly observed in rabbits dosed with antibiotics, confirming that this species is an inappropriate model for studying the teratogenic potential of antibiotics.

Tilmicosin has been tested for reverse mutation in bacteria, forward mutation in cultured mammalian cells, unscheduled DNA synthesis in primary cultures of rat hepatocytes, and in *in vivo* cytogenetic assays. All results were negative, and the Committee concluded that tilmicosin has no genotoxic potential.

The carcinogenic potential of tilmicosin has not been tested. However, a number of factors are relevant in assessing the carcinogenicity of this drug. Toxicity studies with tilmicosin have not resulted in lesions or proliferative changes suggestive of neoplasia, and tilmicosin was uniformly negative in a wide range of genotoxicity assays. Tilmicosin is a macrolide antibiotic, and this class of chemicals is not known to induce neoplasia despite widespread use in humans over many years. The closest structural analogue, tylosin, was reviewed at the thirty-eighth meeting of the Committee (Annex 1, reference 97). In a 2-year feeding study in rats, tylosin administration was associated with an increased incidence of pituitary adenomas in male animals when compared to concurrent controls. New evidence, not considered at the thirty-eighth meeting, indicates that spontaneous pituitary neoplasms occur in male rats at variable rates. Additionally, increased rates occur in animals with higher body weights. In tylosin-treated groups of male rats, the 12-month body weights were somewhat higher than in the control groups and the incidences of pituitary adenomas were comparable to the upper limit of the historical control range. The lower incidence of pituitary neoplasms in the control animals may have been due to the earlier mortality in this group caused by respiratory disease. In light of this information the Committee considered that the concerns of the thirty-eighth meeting over the potential tumorigenicity of tylosin had been satisfactorily addressed and that there was no concern for the carcinogenic potential of tylosin. For the above reasons, the Committee considered that carcinogenicity studies would not be required for tilmicosin.

In assessing the microbiological activity of tilmicosin, the results of a study using germ-free rats colonized with human intestinal microflora was considered the most relevant. At the highest dose administered (0.4 mg/kg bw per day for 5 days) there were no significant alterations in numbers of total anaerobes or enterobacteria in rat faeces.

Reports of accidental human exposure to tilmicosin have identified local reactions indicative of an irritant action following dermal and ocular exposure. Humans accidentally injected with tilmicosin have reported anxiety, sweating, headache and lightheadedness. Changes in the electrocardiogram pattern were observed in only one person. Contact with the buccal mucosa or ingestion has resulted in a range of symptoms including nausea, vomiting, thirst, numbness or burning sensation in the mouth, and headache.

4. EVALUATION

The NOEL from toxicological studies was 4 mg/kg bw per day in a 12-month study in dogs. Treatment of rats colonized with human intestinal microflora with 0.4 mg tilmicosin/kg bw per day produced no significant microbiological effect. The Committee established an ADI of 0-40 μg/kg bw, based on the toxicological NOEL of 4 mg/kg bw per day and a safety factor of 100. An identical ADI would have been established using the data from the study on rats colonized with human intestinal microflora and a safety factor of 10 to account for variability among humans.

5. REFERENCES

Christian, M.S. & Hoberman, A.M. (1989). Reproductive effects of EL-870 administered orally via gavage to Crl:COBS CD(SD) BR rats for two generations, with two litters per generation. Unpublished study No. 112-001 from Argus Research Laboratories. Submitted to WHO by Lilly, Basingstoke, UK.

Cochrane, R.L. & Thomson, T.D. (1990). Toxicology and pharmacology of tilmicosin following administration of subcutaneous and intravenous injections to sheep. Unpublished study No. T5C768908 from Lilly Research Laboratories. Submitted to WHO by Lilly, Basingstoke, UK.

Dearlove, G.H., Hoberman, A.M., & Christian, M.S. (1987). Dosage-range study of EL-870 administered orally via gavage to Crl:COBS CD(SD) BR rats (Pilot study). Unpublished study No. 112-001P from Argus Research Laboratories. Submitted to WHO by Lilly, Basingstoke, UK.

Donoho, A.L. (1988). Comparative metabolism of ^{14}C-tilmicosin in cattle and rats. Unpublished study No. ABC-0395 from Lilly Research Laboratories. Submitted to WHO by Lilly, Basingstoke, UK.

Donoho, A.L. & Kennington, A.S. (1993). Tilmicosin metabolite study with rat excreta. Unpublished study No. T5C759302 from Lilly Research Laboratories. Submitted to WHO by Lilly, Basingstoke, UK.

Donoho, A.L. & Thomson, T.D. (1988). ^{14}C-Tilmicosin balance-excretion study in swine. Unpublished study No. ABC-0409 from Lilly Research Laboratories. Submitted to WHO by Lilly, Basingstoke, UK.

Donoho, A.L., Darby, J.M., Helton, S.L., Sweeney, D.J., Occolowitz, J.L., & Dorman, D.E. (1993). Tilmicosin metabolism study in tissues and excreta of pigs fed 400 ppm ^{14}C-tilmicosin. Unpublished study No. T5C759201 from Lilly Research Laboratories. Submitted to WHO by Lilly, Basingstoke, UK.

Garriott, M.L., Rexroat, M.A., & Jordan, W.H. (1993). The effect of tilmicosin on the induction of reverse mutations in *Escherichia coli* using the Ames test. Unpublished study No. 920825AMS2449 from Lilly Research Laboratories. Submitted to WHO by Lilly, Basingstoke, UK.

Giera, D.D. & Thomson, T.D. (1986). Preliminary ^{14}C EL-870 balance excretion and tissue residue in swine. Unpublished study No. ABC-0305 from Lilly Research Laboratories. Submitted to WHO by Lilly, Basingstoke, UK.

Gries, C.L. & Young, S.S. (1982). Positive correlation of body weight with pituitary tumor incidence in rats. *Fundam. Appl. Toxicol.*, 2, 145-148.

Hawkins, D.R., Elsom, L.F., Dighton, M.H., Kaur, A,. & Cameron, D.M. (1993). The metabolism and residues of ^{14}C-tilmicosin following subcutaneous administration to sheep. Unpublished study No. HRC/LLY 36/930447 from Huntingdon Research Centre. Submitted to WHO by Lilly, Basingstoke, UK.

Jordan, W.H. (1986). The toxicity of compound 177370 (EL-870) given by oral gavage to Fischer 344 rats for two weeks. Unpublished study No. RI5585 from Lilly Research Laboratories. Submitted to WHO by Lilly, Basingstoke, UK.

Jordan, W.H. (1987). The toxicity of EL-870 given orally to Beagle dogs for three months. Unpublished study No. DO8286 from Lilly Research Laboratories. Submitted to WHO by Lilly, Basingstoke, UK.

Jordan, W.H. (1988). The toxicity of tilmicosin given orally to Crl:CD(SD) rats for three months. Unpublished study No. RO9886 from Lilly Research Laboratories. Submitted to WHO by Lilly, Basingstoke, UK.

Jordan, W.H. & Bernhard, N.R. (1989). A one year chronic toxicity study in Beagle dogs given oral doses of tilmicosin. Unpublished study No. DO7187 from Lilly Research Laboratories. Submitted to WHO by Lilly, Basingstoke, UK.

Jordan, W.H. & Higdon, G.L. (1988). A teratology study of tilmicosin (EL-870, compound 177370) administered orally to CD rats. Unpublished study No. RI3387 from Lilly Research Laboratories. Submitted to WHO by Lilly, Basingstoke, UK.

Jordan, W.H. & Ivett, J.L. (1989). Mutagenicity test on tilmicosin in the rat bone marrow cytogenetic assay. Unpublished study No. 10646-0-452 from Hazleton Laboratories. Submitted to WHO by Lilly, Basingstoke, UK.

Jordan, W.H. & Sarazan, R.D. (1991). An acute study of the cardiovascular effects of intravenous Micotil 300 in conscious dogs. Unpublished study No. DV0890 from Lilly Research Laboratories. Submitted to WHO by Lilly, Basingstoke, UK.

Jordan, W.H. & Shoufler, J.R. (1990). The acute toxicity of Micotil 300 administered intramuscularly to Rhesus monkeys. Unpublished studies Nos. PO4089 and PO7689 from Lilly Research Laboratories. Submitted to WHO by Lilly, Basingstoke, UK.

Jordan, W.H., Hill, L.E., & Probst, G.S. (1985). The effect of compound 177370 (EL-870) on the induction of DNA repair synthesis in primary cultures of adult rat hepatocytes. Unpublished studies Nos. 851008UDS2449 and 851015UDS2449 from Lilly Research Laboratories. Submitted to WHO by Lilly, Basingstoke, UK.

Jordan, W.H., McKinley, E.R., Brown, G.E., & Hawkins, D.R. (1986a). The acute toxicity of compound 177370 (EL-870) administered subcutaneously to the ICR mouse. Unpublished studies Nos. M-C-41-85 and M-C-40-85 from Lilly Research Laboratories. Submitted to WHO by Lilly, Basingstoke, UK.

Jordan, W.H., McKinley, E.R., Brown, G.E., & Hawkins, D.R. (1986b). The acute toxicity of compound 177370 (EL-870) administered orally to the Sprague-Dawley rat. Unpublished studies Nos. R-O-40-86 and R-O-39-86 from Lilly Research Laboratories. Submitted to WHO by Lilly, Basingstoke, UK.

Jordan, W.H., McKinley, E.R., Brown, G.E., & Hawkins, D.R. (1986c). The acute toxicity of compound 177370 (EL-870) administered subcutaneously to the Fischer 344 rat. Unpublished studies Nos. R-C-02-86 and R-C-01-86 from Lilly Research Laboratories. Submitted to WHO by Lilly, Basingstoke, UK.

Jordan, W.H., Neal, S.B., & Probst, G.S. (1986d). The effect of compound 177370 (EL-870) on the *in vivo* induction of sister chromatid exchange in bone marrow of Chinese hamsters. Unpublished study No. 851014SCE2449 from Lilly Research Laboratories. Submitted to WHO by Lilly, Basingstoke, UK.

Jordan, W.H., Oberly, T.J., Bewsey, B.J., & Probst, G.S. (1986e). The effect of compound 177370 (EL-870) on the induction of forward mutation at the thymidine kinase locus of L5178Y mouse lymphoma cells. Unpublished studies Nos. 851106MLA2449 and 851113MLA2449 from Lilly Research Laboratories. Submitted to WHO by Lilly, Basingstoke, UK.

Jordan, W.H., Rexroat, M.A., & Probst,G.S. (1986f). The effect of compound 177370 (EL-870) on the induction of reverse mutations in *Salmonella typhimurium* using the Ames test. Unpublished studies Nos. 850930AMS2449, 851028AMS2449 and 851111AMS2449 from Lilly Research Laboratories. Submitted to WHO by Lilly, Basingstoke, UK.

Jordan, W.H., Markey, T.F., Oakley, L.M., & Torrence, T.L. (1987). The acute dermal and inhalation toxicity and ocular and dermal irritation of tilmicosin technical. Unpublished studies Nos. B-D-45-86, B-E-39-86 and R-H-037-86 from Lilly Research Laboratories. Submitted to WHO by Lilly, Bsasingstoke, UK.

Jordan, W.H., Williams, P.D., & Turk, J.A. (1988). Cardiovascular effects of tilmicosin administered intravenously in anaesthetised beagle dogs. Unpublished study No. PC8806 from Lilly Research Laboratories. Submitted to WHO by Lilly, Basingstoke, UK.

Jordan, W.H., Brunny, J.D., & Garriott, M.L. (1989a). The effect of tilmicosin on the *in vivo* induction of sister chromatid exchange in bone marrow of Chinese hamsters. Unpublished study No. 881114SCE2449 from Lilly Research Laboratories. Submitted to WHO by Lilly, Basingstoke, UK.

Jordan, W.H., Gardner, J.B., & Weaver, D.E. (1989b). An intracutaneous sensitisation study in albino guinea pigs with tilmicosin. Unpublished study No. GO1888 from Lilly Research Laboratories. Submitted to WHO by Lilly, Basingstoke, UK.

Jordan, W.H., Richardson, K.K., & Oberly, T.J. (1989c). The effect of tilmicosin on the induction of forward mutation at the HGPRT$^+$ locus of Chinese hamster ovary cells. Unpublished studies Nos. 881130CHT2449 and 890111CHO2449 from Lilly Research Laboratories. Submitted to WHO by Lilly, Basingstoke, UK.

Jordan, W.H., Williams, P.D., & Helton, D.R. (1989d). CNS behavioural effects of tilmicosin administered orally in the male CD-1 mouse. Unpublished study No. PN8805 from Lilly Research Laboratories. Submitted to WHO by Lilly, Basingstoke, UK.

Jordan, W.H., Williams, P.D., & Colbert, W.E. (1990). *In vitro* studies of tilmicosin in the cardiac muscle of Hartley Albino guinea pigs. Unpublished study No. PM8946 from Lilly Research Laboratories. Submitted to WHO by Lilly, Basingstoke, UK.

Jordan, W.H., Wolff, R.K., & Carlson, K.H. (1991). A subchronic inhalation toxicity study of tilmicosin in Beagle dogs. Unpublished study No. DO5589 from Lilly Research Laboratories. Submitted to WHO by Lilly, Basingstoke, UK.

McGrath, J.P., Hamelink, J.L., & Lochmueller, C.A. (1988a). A study of the immune response in CD-1 mice treated orally with tilmicosin. Unpublished study No. MI4288 from Lilly Research Laboratories. Submitted to WHO by Lilly, Basingstoke, UK.

McGrath, J.P., Hamelink, J.L,. & Stephenson, R.O. (1988b). A study of the acute effects of urine and electrolyte excretion of tilmicosin administered orally to female Crl:CD(SD) rats. Unpublished study No. RI8688 from Lilly Research Laboratories. Submitted to WHO by Lilly, Basingstoke, UK.

McGuigan, M.A. (1994). Human exposures to tilmicosin (Micotil). *Vet. Hum. Toxicol.*, **36**(4), 306-308.

McLaren, I.M. (1994). Determination of the minimum inhibitory concentration of tilmicosin for *Lactobacillus* and *Propionibacterium* Sp. Unpublished study from Central Veterinary Laboratory, UK. Submitted to WHO by Lilly, Basingstoke, UK.

Montanio, C.D. & Dart, R.C. (1993). Micotil 300. Human exposures May 1992-May 1993. Unpublished report. Submitted to WHO by Lilly, Basingstoke, UK.

Noda, A. (1993). Teratogenicity study of EL-870 (tilmicosin aqueous) in rabbits by gavage. Unpublished study No. 91-001 from Research Institute for Animal Science in Biochemistry and Toxicology, Japan. Submitted to WHO by Lilly, Basingstoke, UK.

Piroozi, K.S., Keaton, M.J., & Jordan, W.H. (1993). The acute toxicity of tilmicosin administered orally to Fischer 344 rats. Unpublished study No. R21493 from Lilly Research Laboratories. Submitted to WHO by Lilly, Basingstoke, UK.

Rumney,C.J. (1993). Microbiological end-point determination for two antibiotics. Unpublished study No. 1164/3/93 from Bibra Toxicology International, UK. Submitted to WHO by Lilly, Basingstoke, UK.

Sarazan, R.D., Main, B.W., & Jordan, W.H. (1993). Cardiovascular effects of tilmicosin alone, or in combination with propranolol or dobutamine, in conscious unrestrained dogs. Unpublished study No. DV1591 from Lilly Research Laboratories. Submitted to WHO by Lilly, Basingstoke, UK.

Scott, R.J.D., Morgan, J.M., Pether, J.V.S., & Gosling, P. (1993). Determination of the minimum inhibitory concentrations of tilmicosin for human isolates of *Bifidobacterium* and *Peptostreptococcus*. Unpublished study No. SMS/9301 from Southern Microbiological Services, UK. Submitted to WHO by Lilly, Basingstoke, UK.

Williams, P.D., Jordan, W.H., & Colbert, W.E. (1988). *In vitro* studies of tilmicosin (EL-870, Compound 177370) in smooth and cardiac muscle. Unpublished study No. PM8808 from Lilly Research Laboratories. Submitted to WHO by Lilly, Basingstoke, UK.

INSECTICIDES

CYPERMETHRIN & α-CYPERMETHRIN

First draft prepared by
Mrs Ir. M.E.J. Pronk,
Dr G.J.A. Speijers,
Mrs M.F.A. Wouters
Toxicology Advisory Centre
National Institute of Public Health and Environmental Protection
Bilthoven, Netherlands

Dr L. Ritter
Canadian Network of Toxicology Centres
University of Guelph
Guelph, Ontario, Canada

1.   EXPLANATION

Cypermethrin and α-cypermethrin are highly active synthetic pyrethroid insecticides, effective against a wide range of pests in agriculture, public health and animal husbandry. Cypermethrin has been widely used throughout the world since the late 1970s while α-cypermethrin has been available commercially since the mid 1980s.

Cypermethrin and α-cypermethrin are neuropoisons acting on the axons in the peripheral and central nervous system by interacting with sodium channels in mammals and insects.

Cypermethrin and α-cypermethrin have not been previously evaluated by the Committee. However, cypermethrin was reviewed by the Joint FAO/WHO Meeting on Pesticide Residues (JMPR) in 1979 and 1981 (FAO, 1980, 1982); an ADI of 0-0.05 mg/kg bw was established at the 1981 meeting. Studies reviewed in 1979 and 1981, which were not available at the present meeting were considered in this evaluation on the basis of the JMPR summaries.

Cypermethrin consists of eight isomers, four cis and four trans isomers, the cis isomers being the more biologically active. Depending on the manufacturing source, the cis:trans ratio varies from 40:60 to 80:20. In the studies submitted for evaluation the cis:trans ratio was in the range of 48:52 to 50:50. The purity varied between 92-95.1%.

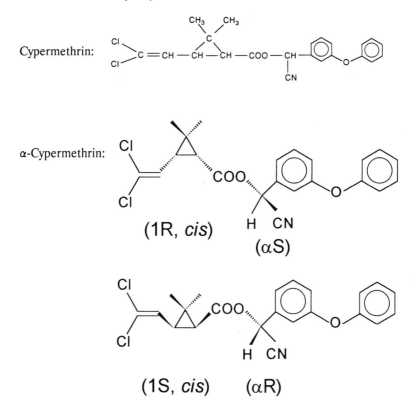

α-Cypermethrin contains more than 90% of the insecticidally most active enantiomer pair of the four cis isomers of cypermethrin as a racemic mixture.

2.   BIOLOGICAL DATA

2.1   Biochemical aspects

2.1.1   Absorption, distribution and excretion

According to the JMPR (FAO, 1980, 1982) cypermethrin is readily absorbed and rapidly eliminated via urine and faeces of mice, rats, dogs, sheep and cows. Absorption from the gastrointestinal tract is more rapid with the trans isomer than with the cis isomer. The highest mean concentrations are found in body fat, liver, kidney, muscle, skin and milk. The clearance rate from adipose tissue is slow and the elimination half-life in rats and mice ranges from 10 to 20 and from 20 to 30 days, respectively. The data suggest a potential for bioaccumulation in the body following continuous exposure. With respect to the chemical and especially the isomeric complexity of the molecule, the metabolic profile due to all of its isomers is extremely complex. Cypermethrin is readily cleaved at the ester linkage to produce the cyclopropane carboxylic acid and a 3-phenoxybenzoyl moiety that is further metabolised by oxidation at the 4'position. The resultant phenol is almost totally conjugated with sulfate. The 4'-hydroxy sulfate forms the major aryl metabolite (16%), and 3-phenoxybenzoic acid is the second most important (5%). The other identified aryl metabolites are 3-(4-hydroxyphenoxy)benzoic acid (1%) and the glycine conjugate, N-(3-phenoxybenzoyl) glycine (1%).

2.1.1.1   Rats

Eight female Wistar rats were orally dosed with 2.5 mg ^{14}C-benzyl-labelled cis-cypermethrin/kg bw. They were killed in groups of two at 8, 14, 25 and 42 days after dosing, and the radioactivity in fat, liver and kidney was measured. Selected fat samples were analysed for parent compound. Between days 14 and 42, radioactivity was eliminated from the fat with a half-life of 20-25 days. Residues in liver and kidney were 30 to 40 times lower than those in fat, but were eliminated at a similar rate. In two pooled fat samples taken 8 and 24 days after dosing, 90-100% of the radioactivity was in the form of the parent compound (Crawford & Hutson, 1978).

2.1.1.2   Chickens

Six Warren laying hens were treated daily for 14 days with 10 mg
[14]C-phenoxy-labelled cypermethrin/kg food (equivalent to 1.25 mg/kg bw
per day) in gelatin capsules in the diet. Eggs and excreta were collected
daily. The hens were killed 4.5 hours after the last dose, and liver, muscle
and fat samples were collected. Of the total radioactivity, 95.2% was
recovered. Radioactivity in eggs plateaued 8 days after the start of dosing
and reached 0.05 mg/kg. Most radioactivity was found in the yolk and was
a mixture of parent compound and material that was closely associated with
neutral lipids and phosphatidyl cholines. In tissues, highest radioactivity was
found in the liver (0.37 μg/g). This radioactivity was composed of parent
compound (0.05 μg/g) and a mixture of very polar metabolites that were
not hydrolysed to significant amounts of 3-phenoxybenzoic acid or its
4-hydroxy derivative. Fat contained 0.08 μg/g and about 60% of the residue
was present as parent compound. Residues in muscle (0.01-0.02 μg/g) were
too low for characterization (Hutson & Stoydin, 1987).

2.1.1.3   Sheep

Two male sheep were each treated dermally (single dose) with a
mixture of 962 mg ^{14}C-cypermethrin (both cyclopropyl- and benzyl-labelled),
equivalent to 21.9 mg/kg bw in acetone. After 24 hours (sheep 1) and 6
days (sheep 2) the animals were killed, and radioactivity in fat, muscle, liver,
kidney and skin (application area) was measured. Urine and faeces were
collected up to the time the sheep were killed. One sheep treated with
acetone served as a control. Another sheep was treated orally (single dose)
with 177 mg ^{14}C-cypermethrin mixture in a gelatin capsule (equivalent to 3.9
mg/kg bw). Urine and faeces were collected for 2 days and tissues were
assayed. When applied dermally radioactivity was slowly absorbed and
eliminated. Less than 0.5% was excreted in urine within 24 hours and only
2% over a 6-day period. Faecal elimination was also slow, 0.5% of the dose
being eliminated in 6 days. About 30% of the applied radioactivity was
recovered from the application areas of both sheep. Very little radioactivity
was found in the tissues. In liver, kidney, renal fat and subcutaneous fat
(other than the application area), residues ranged from 0.09 to 0.3 μg/g and
muscle samples contained 0.03 to 0.06 μg/g.
Following oral treatment, the elimination of radioactivity was rapid,
61% of the administered radioactivity being eliminated within 48 hours.
Urinary elimination comprised 41% of the dose and faecal elimination
20.5%. Residues found in tissues were comparable to those found after
dermal application. Most of the radioactivity in the fat samples of all sheep
was found to be parent compound. In muscle, liver and kidney samples, only
a small percentage of the total tissue radioactivity was identified as parent
compound (Crawford & Hutson, 1977).

2.1.1.4   Cattle

a)   Cypermethrin

Two lactating Friesian cows were fed twice daily a diet containing 0.2 mg ^{14}C-benzyl-labelled cypermethrin/kg feed. Cow I was treated for 20 days and cow II for 21 days. A control cow was maintained under identical conditions. The radioactivity in milk, urine and faeces was measured daily and after 20-21 days of dosing the cows were sacrificed and blood, major organs and tissues were analysed for radioactivity. Total recovery of the radioactivity was 97.8%. The major route of excretion was via the urine (54%) and faeces (43%). Milk contained 0.5% of the dose. The residues in muscle, blood and brain were less than 0.005 μg/g. Liver contained 0.006 μg/g, kidney 0.004 μg/g, bile 0.027 μg/g, renal fat 0.011 μg/g and subcutaneous fat 0.009 μg/g.

The urinary metabolites were tentatively identified as N-(3-phenoxybenzoyl)glutamic acid, the major metabolite, and 3-(4-hydroxyphenoxy) benzoic acid O-sulphate, the minor metabolite (ratio 4:1). In faeces 36% was eliminated as unchanged parent compound (Hutson & Stoydin, 1976).

Two lactating Friesian cows were fed twice daily for 7 days a diet containing 5 mg ^{14}C-cypermethrin (cyclopropyl-labelled)/kg feed. A third cow was treated similarly with ^{14}C-benzyl-labelled cypermethrin. A control cow was maintained under identical conditions. The radioactivity in the milk, urine and faeces was measured daily. After 7 days of dosing the cows were killed and their blood, major organs and tissues were analysed for radioactivity. Total recovery of radioactivity was 92.4% for the cows dosed with cyclopropyl-labelled cypermethrin and 89.6% for the cow dosed with benzyl-labelled cypermethrin. The major route of excretion was similar for both label compounds. Urinary excretion accounted for about 49% of the dose while faecal excretion accounted for about 38%. An equilibrium between ingestion and excretion was reached after 3-4 days. Very low levels of radioactivity were determined in milk (0.003-0.013 μg/g), tissues (muscle, blood, brain 0.07 μg/g, liver, kidney 0.13 μg/g and renal and subcutaneous fat 0.10 μg/g). The urinary metabolites included the glutamic acid conjugate of 3-phenoxybenzoic acid (68%), 3-phenoxy-benzoylglycine (16%) and 3-phenoxybenzoic acid (9%). 3-(4-Hydroxyphenoxy)benzoic acid and its O-sulfate conjugate appeared to be present in only small amounts (1%) (Crawford, 1978).

One lactating cow was administered a diet containing 10 mg ^{14}C-benzyl-labelled cypermethrin/kg feed twice daily for 7 days. Milk, urine and faeces were collected daily for radioassay. The cow was killed 16 hours after the last dose and samples of fat, muscle, liver and kidney were analysed. An untreated cow was held as control. Total recovery was 93%. Radioactivity was rapidly eliminated in equal proportions in the urine and faeces. Radioactivity in milk was < 0.2%. Analysis of milk revealed that the radioactivity was due to unchanged cypermethrin, which was located mostly in the lipophilic components (cream or butterfat). Radioactivity in tissues was generally in the order: liver (0.21 μg/g) > kidney (0.11 μg/g) > fat (0.1 μg/g) > blood (0.04 μg/g) > muscle (0.01 μg/g). The residue in fat was largely unchanged cypermethrin. Radioactivity in liver and kidney was due mainly to N-(3-phenoxybenzoyl)glutamic acid. The liver and kidney metabolites were hydrolysed in hot acid to 3-phenoxybenzoic acid and its 4'-hydroxy derivative (Croucher et al., 1980).

b)   α-Cypermethrin

One lactating cow received orally 125 mg ^{14}C-α-cypermethrin (benzyl labelled)/dose twice daily for 4 consecutive days (target dose 250 mg/day). Another cow received the same dose of unlabelled α-cypermethrin. (The overall calculated daily dietary concentrations were 19 and 14 mg/kg, respectively). The animals were killed 6 hours after the last dose. Urine, faeces, milk, kidneys, liver, fat and muscle were analysed for radioactivity.

The major route of excretion of radioactivity was via the faeces, accounting for 34% of the total administered dose. Urinary excretion accounted for 23% of the dose and milk less than 1%. Total radioactive residues in milk accounted for 0.014 mg/kg on day 2 and rose to 0.2 mg/kg by day 4 of dosing. The major proportion (93%) of the milk residues was confined to the cream fraction. Residues in tissues were highest in liver, renal fat, omental fat, subcutaneous fat and kidney (0.56, 0.48, 0.43, 0.39 and 0.22 mg/kg, respectively). Plasma contained 0.08 mg/kg and muscle samples contained < 0.03 mg/kg residues.

The liver and kidney contained a range of components. The liver extract contained at least eight metabolites with a broad range of polarities. One component (16%) had similar chromatographic properties to α-cypermethrin. The kidney extract contained at least nine metabolites with a broad range of polarities, one component (20% of the profile) had similar chromatographic properties to α-cypermethrin. Muscle, fat and milk contained mainly a single component (muscle 85%, fat 91% and milk 97% of the exact profile), which in each case had similar chromatographic properties to α-cypermethrin. Urinary metabolites were analysed using HPLC. The two major components (44% and 20%) had identical chromatographic properties to N-(3-phenoxybenzoyl)glutamic acid and N-(3-phenoxybenzoyl)glycine, respectively. A minor component (3%) had identical chromatographic properties to 3-phenoxybenzoic acid (Dunsire & Gifford, 1993; Morrison & Richardson, 1994).

2.1.2   Effects on liver enzymes

Six Wistar rats/sex were randomly selected from each of the control and 1000 mg cypermethrin/kg food group at termination of a 2-year feeding study. The activity of hepatic *p*-nitroanisole *o*-demethylase (PNOD) was determined in each of these rats. In male rats treated with cypermethrin, PNOD activity, expressed as per gram of liver, was significantly increased (38%). The total PNOD activity of liver was also increased (30%), but, owing to great inter-animal variation, statistical significance was not reached. Treated females exhibited increases in PNOD activity when expressed per gram of liver (21%) and per whole liver (39%), but only the latter was significant (Potter & McAusland, 1980).

2.2   Toxicological data

2.2.1   Acute toxicity

Acute toxicity studies of cypermethrin and α-cypermethrin are summarized in Tables 1 and 2.

Cypermethrin-induced signs of toxicity were typical of cyano-containing pyrethroid intoxication. After oral administration signs of intoxication included sedation, ataxia, splayed gait, tip-toe walking, with occasional tremors and convulsions. These signs of toxicity appeared within a few hours following dosing, and survivors recovered within 3 days. The acute toxicity after dermal administration is of a low order (Coombs *et al.*, 1976).

α-Cypermethrin-induced clinical signs, like those of cypermethrin, are typical of cyano-containing pyrethroid. The observed signs included ataxia, abasia, gait abnormalities, choreoathetosis, tip-toe walking, and increased salivation, lacrimation, piloerection, tremor and clonic convulsions. Surviving animals recovered within 7 days (Rose, 1982b, 1983).

2.2.2   Short-term toxicity studies

2.2.2.1   Mice

Groups of eight mice (CD-1)/sex received diets containing 0, 200, 400, 800, 1200 or 1600 mg α-cypermethrin/kg feed for 29 days. One male fed 1600 mg/kg and one female fed 1200 mg/kg were killed in a moribund condition. These animals showed neurological disturbances. Mice at 1200 and 1600 mg/kg developed ungroomed coats, ataxia/abnormal gait, over-activity or hunched posture. At 800 mg/kg some animals had ungroomed coats and abnormal gait. Body weight (gain) was dose-relatedly decreased in mice at 1200 and 1600 mg/kg and in females at 800 mg/kg. A similar, but less marked effect was apparent in males receiving 800 mg/kg and females receiving 400 mg/kg. Food consumption was lowered in rats at 1200 and 1600 mg/kg during the first 2 weeks. A depression in lymphocyte numbers was noted in males receiving 1600 mg/kg. ALAT and ASAT levels were increased in males at 1600 mg/kg. Urea levels were increased in all

Table 1. Acute toxicity of cypermethrin in animals

Species[1]	Route	Purity[2]	LD$_{50}$ mg/kg bw	Remark	Reference
Mouse	oral	98.1%	88	5% in corn oil	Rose, 1982a
Mouse	oral	?	82	5% in corn oil	Coombs *et al.*, 1976
Mouse	oral	98.1%	1126	40% in DMSO	Rose, 1982a
Mouse	oral	98.1%	657	50% aq. susp.	Rose, 1982a
Mouse	percutan.	98.1%	> 100	5% in corn oil	Rose, 1982a
Rat	oral	?	251	5% in corn oil	Coombs *et al.*, 1976
Rat	oral	98.1%	4000	40% in DMSO	Rose, 1982a
Rat	oral	98.1%	3423	50% aq. susp.	Rose, 1982a
Rat	oral	?	> 2000	techn. conc.	Price, 1984
Rat	percutan.	?	> 2000	techn. conc.	Price, 1984
Rat	percutan.	?	> 1600	40% in xylene	Coombs *et al.*, 1976
Rat	i.p.	?	~ 2600	40% in corn oil	Coombs *et al.*, 1976
Hamster	oral	?	203	5% in corn oil	Coombs *et al.*, 1976
Guinea-pig	oral	?	~ 500	20% in corn oil	Coombs *et al.*, 1976

[1] Both male and female animals were used in all of the studies
[2] cis:trans isomer ratio 51:49; ? - unknown

dosed males and females, without clear dose relation. Plasma albumin and A/G ratio were decreased in males at 1600 mg/kg. Relative kidney weight was increased in males at 800, 1200 and 1600 mg/kg. Relative lung weight was increased in males at 1600 mg/kg. No macroscopic or microscopic effects were seen. The NOEL was 400 mg/kg feed, equal to 57 mg/kg bw per day (Green, 1993).

Groups of 12 mice (CD-1)/sex received diets containing 0, 50, 250 or 1000 mg α-cypermethrin/kg feed for 13 weeks. Four males at

Table 2. Acute toxicity of α-cypermethrin in animals

Species[1]	Route	Purity (%)	LD_{50} mg/kg bw or mg/m³	Remark	Reference
Mouse	oral	96.6	35	5% in corn oil	Rose, 1982b
Mouse	oral	96.6	762	40% in DMSO	Rose, 1982b
Mouse	oral	96.6	798	50% aq. suspension	Rose, 1982b
Mouse	percut.	96.6	> 100	5% in corn oil	Rose, 1982b
Rat	oral	techn.	64	50% aq. CMC	Gardner, 1993
Rat	oral	96.6	4000	40% in DMSO	Rose, 1982b
Rat	oral	96.6	> 5000	50% aq. suspension	Rose, 1982b
Rat	dermal	techn.	> 2000	undiluted	Gardner, 1993
Rat	inhal	95.6	1590	as aerosol (MMAD 9 μm)	Jackson, 1993

[1] Both male and female animals were used in all of the studies

1000 mg/kg died during week 12 probably due to stress of treatment and refusal of food and water. The relative kidney weights of these animals appeared to be slightly higher. One mouse at 250 mg/kg died. Clinical signs included thin build, ungroomed coat, hair loss and encrustations of the dorsal body surface in animals receiving 1000 mg/kg, ungroomed haircoat in mice at 250 mg/kg, hair loss and encrustations in males at 250 mg/kg and hair loss in two males at 50 mg/kg. Body weight gain was markedly lower in mice at 1000 mg/kg and lower in mice at 250 mg/kg. Food consumption was slightly higher in mice at 1000 mg/kg. Overall food conversion efficiency of animals receiving 1000 mg/kg (and to a lesser extent 250 mg/kg) was lower than that of controls. In males at 1000 mg/kg, PVC, Hb, RBC, total WBC and leucocyte counts were decreased. ASAT was increased in a dose-dependent manner in males at 250 and 1000 mg/kg, and glucose was decreased in males at 1000 mg/kg. Serum AP was increased in females at 1000 mg/kg. Urinary specific gravity was increased in mice at the highest dose level. In males at 1000 mg/kg relative brain, adrenal, heart, kidney, liver, spleen, lung and testes weight were increased. In females relative liver weight was increased at 250 and 1000 mg/kg and relative brain and spleen weight at 1000 mg/kg. Two males and eleven females treated at 1000 mg/kg were considered to be emaciated at necropsy. No histological changes were observed. Owing to the hair loss seen at 50 mg/kg, a NOEL was not identified (Amyes *et al.*, 1994).

2.2.2.2   Rats

a)   Cypermethrin

In a 5-week feeding study, groups of six rats/sex (Charles River) received diets containing 0, 25, 100, 250, 750 or 1500 mg cypermethrin/kg feed. No effects were found in haematological, macroscopic or microscopic examinations. In the highest dose group, rats developed piloerection, nervousness and incoordinated movement from week 2 onwards. Body weight gain, food intake and terminal body weight were all reduced for both sexes at 1500 mg/kg. Relative liver weight and blood urea levels were increased at the highest dose level (only a summary available) (Coombs *et al.*, 1976).

Groups of 12 rats/sex (Wistar) were fed diets containing 0 (24 rats/sex), 25, 100, 400 or 1600 mg cypermethrin/kg feed for 91-95 days. There were no deaths, no clinical signs and no macroscopic or microscopic changes. In the highest dose group a reduced body weight gain was observed. During the first week reduced food consumption was observed in both sexes at 1600 mg/kg. This could be attributed to the palatability of the diet. During week 13, females at 1600 mg/kg had reduced food intake. Males at 1600 mg/kg showed decreases in Hb, MCV and eosinophil counts, and an increase in prothrombin time. In males at 400 mg/kg there was also a decrease in eosinophil numbers. The mean plasma urea concentration in the 1600 mg/kg groups was increased. Relative liver weight was increased in males and females at 1600 mg/kg and in males at 400 mg/kg. Relative kidney weight was increased in males at 1600 mg/kg. The NOEL was 100 mg/kg feed, equivalent to 5 mg/kg bw per day (Pickering, 1981).

b)   α-Cypermethrin

Groups of 10 Wistar rats/sex received diets containing 0 (20 rats/sex), 20, 100, 200, 400 or 800 mg α-cypermethrin/kg feed for 5 weeks. Observations included mortality, clinical signs, body weight and food consumption, haematology, clinical chemistry, urinalysis, organ weight, and macroscopic and microscopic lesions.

One male at 800 mg/kg died unrelated to treatment and two males at 800 mg/kg were killed, owing to signs of severe neurological disturbance. Both sexes at 800 mg/kg produced abnormal gait and increased sensitivity to noise. Abnormal gait was seen in one male at 400 mg/kg. The mean body weights and food intake of both males and females fed 400 or 800 mg/kg were significantly lower than those of controls. In both males and females at 800 mg/kg and in males at 400 mg/kg prothrombin time was increased. In males at 800 mg/kg the kaolin-cephalin coagulation time was increased and the percentage of polymorphic neutrophils was increased at 800 mg/kg. In females platelet count, total leucocyte count and absolute values of polymorphic neutrophils and lymphocytes were increased, but the Hb and haematocrit values were decreased. The MCV was decreased at 200, 400 and 800 mg/kg but not in a dose-related manner. Relative brain, liver and kidney weights were increased in males fed 800 mg/kg and relative brain and liver weights were increased in males fed 400 mg/kg. In females at 800 mg/kg relative brain, liver and kidney weights were increased. One male at 800 mg/kg that was removed from the study showed scanty axonal lesions of the sciatic nerves. The NOEL was 200 mg/kg feed, equivalent to 10 mg/kg bw per day (Thorpe, 1982).

In a range finding study groups of five male and five female rats were fed diets containing α-cypermethrin at concentrations of 0, 50, 200, 800 or 1200 mg/kg feed for up to 6 weeks. No effects were found on food consumption, haematology and clinical chemistry and microscopy. All rats at 1200 mg/kg and all males at 800 mg/kg were killed during weeks 2 to 4 because of severe clinical signs. These signs included high stepping, splayed gait, abasia and hypersensitivity. Cachexia was seen in extreme cases. There were no significant changes in fore or hind limb grip strength or hind limb landing foot splay. Female rats fed 800 mg/kg had a lower mean body weight and food intake compared to controls. These females also had a lower leucocyte count. Microscopic examination revealed lymphocytolysis and lymphocyte depletion of the cortical region of the thymus in males fed 800 mg/kg and males and females at 1200 mg/kg (Fokkema, 1994a).

Groups of 30 male and 30 female Wistar rats received diets containing 0 (60 rats/sex), 20, 60, 180 or 540 mg α-cypermethrin/kg feed for 90 days. After 6 weeks 10 rats/sex (controls 20/sex) were sacrificed for interim examinations. Three males at 540 mg/kg developed an abnormal gait consisting of splayed hind limbs. Skin sores and fur loss were observed in all males with the highest incidence at 540 mg/kg and in females at 0 and 540 mg/kg. Two males, one from controls and one at 540 mg/kg were removed because of severe skin lesions. Reduced body weight and reduced food consumption were seen in rats at 540 mg/kg. During the second part of the study, body weight gain was also reduced in males at 60 and 180 mg/kg, but not in a clearly dose-related manner. Food consumption was lower at 60 mg/kg than at 180 mg/kg. At termination the Hb value was decreased in males and females at 540 mg/kg. MCV and MCHC were decreased in females at 540 mg/kg. Platelet counts were increased in males and females at 540 mg/kg. The number of lymphocytes was increased and the number of eosinophils was decreased in males at 540 mg/kg. Urea concentration was increased in females at 540 mg/kg. AP was decreased in females at 180 and 540 mg/kg. Urinary volume was decreased in females at 540 mg/kg and specific gravity was increased in males and females at 540 mg/kg. In females fed 540 mg/kg, relative spleen, heart and brain weights were increased. In both sexes at 540 mg/kg relative kidney weight was increased and relative liver weight was increased in both sexes at 180 and 540 mg/kg. Relative testis weight was increased in males at 540 mg/kg. Histopathological investigations showed a solitary form of axonal degeneration, affecting the fibre only, in the sciatic nerve, in two males fed 540 mg/kg, but there were no clinical signs of toxicity. The NOEL was 60 mg/kg feed, equivalent to 3 mg/kg bw per day (Clark, 1982).

2.2.2.3   Rabbits

Occluded dermal applications of 0, 2, 20 or 200 mg cypermethrin/kg bw in PEG 300 were made to abraded and intact skin of groups of 10 male and 10 female NZW rabbits/sex. The applications were made for 6 h/day, 5 days/week for 3 weeks. No effects were observed on haematology, clinical chemistry or following macroscopic or microscopic examination. Slight to severe skin irritation was observed in rabbits at 200 mg/kg bw and slight to moderate skin irritation was observed in rabbits receiving 2 and 20 mg/kg bw. Decreased food consumption and body weight gain was observed in rabbits treated with 200 mg/kg bw. Absolute and relative gonad weights were reduced in male rabbits at 200 mg/kg bw. The NOEL was 20 mg/kg bw (Henderson & Parkinson, 1981).

2.2.2.4   Dogs

a)   Cypermethrin

In a 5-week feeding study, groups of three beagle dogs/sex received diets containing 0, 15, 150 or 1500 mg cypermethrin/kg feed. No effects were found on body weight, food consumption, ophthalmoscopy, haematology, clinical chemistry, macroscopy, liver weight or following microscopic examination. In the highest dose group dogs developed signs of intoxication, including apprehension, diarrhoea and vomiting, licking and chewing of the paws, whole body tremors, a stiff exaggerated hind leg gait, ataxia, inappetence, and decreased body weight gain. Relative thyroid weight and blood urea levels were increased and blood glucose levels were decreased (only a summary available) (Coombs et al., 1976).

Groups of beagle dogs (four/sex) were fed cypermethrin (purity 98%) in the diet at concentrations of 0, 5, 50, 500 or 1500 mg/kg feed for 13 weeks. No mortality occurred. Dogs at 1500 mg/kg developed diarrhoea, anorexia and tremors as well as ataxia, incoordination and hyperaesthesia. Also a reduced food intake and body weight gain was seen. Due to the clinical signs two dogs/sex at 1500 mg/kg had to be killed. Minor variations in haematology were observed; the kaolin-cephalin clotting time was consistently lower throughout the study in female dogs at 500 mg/kg. However, because of the lack of a dose-response relationship, the variability of this parameter throughout the study, and the fact that there were no changes in the prothrombin time, this finding is not considered to be relevant. No effects were seen on clinical chemistry parameters or on organ weights. Microscopic examination of tissues and organs revealed a non-specific focal bronchopneumonia in the lungs of the dogs at 1500 mg/kg. A pink colour of the optic disc was noted in ophthalmological examinations. The NOEL was 500 mg/kg feed, equivalent to 12.5 mg/kg bw per day (Buckwell & Butterworth, 1977).

A further experiment was undertaken to determine the cause of the pink discoloration, which occurred in all groups. Cypermethrin was fed to two groups of three male dogs at concentrations of 0 or 500 mg/kg food for 13 weeks. Specific ophthalmological examinations were made to evaluate the degree of coloration of the optic disc. At the end of 13 weeks, there were no consistent differences between the colour of the optic discs of the treated dogs and the controls (Buckwell & Butterworth, 1977).

b)   α-Cypermethrin

This study was conducted in two parts.
In the first part one male and one female beagle dog were fed α-cypermethrin in the diet at concentrations of 200 mg/kg feed for 7 days, 400 mg/kg feed for 2 days and 300 mg/kg feed for 7 days. Due to severe intoxication of the animals receiving 400 mg/kg, dosing was discontinued after 2 days. The animals were fed the control diet and in week 3 treatment was commenced at 300 mg/kg. After the third week the dogs were killed.
Clinical signs including ataxia, body tremors, subdued behaviour, head nodding, food regurgitation, diminished response to stimuli and inflammation of gums and tongue were obtained when dogs were dosed with 300 and 400 mg/kg. Body weight loss was observed at 300 mg/kg. Examination of the cellular composition of blood and the chemical components of plasma showed no abnormalities, and no macroscopic changes were observed.
In the second part one male and one female beagle dog were fed α-cypermethrin in the diet at 300 mg/kg food for 3 days (male dog) or 4 days (female dog), and then 250 mg/kg food for 7 days. At 300 mg/kg the same effects were obtained as above. At 250 mg/kg only the female dog developed the clinical signs shown by the dogs treated with 300 and 400 mg/kg (Greenough & Goburdhun, 1984).

Groups of four beagle dogs/sex received diets containing dose levels of 0, 30, 90 or 270 (six dogs/sex) mg α-cypermethrin/kg feed for 13 weeks.

No effects on mortality, body weight, food consumption, ophthalmoscopy, haematology, clinical chemistry, macroscopy, organ weights or microscopy were seen. One female at 270 mg/kg was killed, because of severe head and body tremors, ataxia, poor limb coordination, inflamed gums and elevated temperature. All dogs at 270 mg/kg developed marked clinical signs, including body tremors, head nodding, lip licking, subduedness, ataxia, agitation and high stepping gait. The NOEL was 90 mg/kg, equivalent to 2.25 mg/kg bw per day (Greenough *et al.*, 1984).

Four groups of beagle dogs (four/sex) received via the diet 0, 60, 120 or 240 mg α-cypermethrin/kg feed daily for 52 weeks. No effects were observed on body weight, food consumption, ophthalmoscopy, haematology, clinical chemistry, urinalysis or organ weights, and no changes were observed in macroscopic or microscopic examinations. Two males at 240 mg/kg developed skin reddening on the tail. Abdominal skin reddening and alopecia were seen in another male at 240 mg/kg and one female at 120 mg/kg. The tail reddening caused obvious irritation and resulted in ulceration and necrosis in one male leading to amputation of part of the tail. The NOEL was 60 mg/kg feed, equivalent to 1.5 mg/kg bw per day (Dean & Jackson, 1995).

2.2.3   Long-term toxicity/carcinogenicity studies

2.2.3.1   Mice

Groups of mice (70/sex, SPF-Swiss-derived) received diets containing 0 (2 groups), 100, 400 or 1600 mg cypermethrin/kg feed for up to 101 weeks. Ten mice/sex were killed after 52 weeks for interim necropsy. Observations included mortality, clinical signs, body weight gain, food consumption, haematology, clinical chemistry, organ weights, and macroscopy and microscopy. Body weight gain of both males and females at 1600 mg/kg was reduced when compared to the combined control groups. Several haematological changes, consistent with mild anaemia, were found in the 1600 mg/kg group at the interim kill, but not at termination. At interim kill and at termination, an increase in thrombocytosis and absolute and relative liver weight was seen in males at 1600 mg/kg. An increase in the incidence of benign alveologenic tumours was observed in females at 1600 mg/kg, but was within historical control incidence. In this study the NOEL was 400 mg/kg feed, equal to 57 mg/kg bw per day (Lindsay *et al.*, 1982).

2.2.3.2   Rats

Groups of 48 Wistar rats/sex received diets containing 1, 10, 100 or 1000 mg cypermethrin/kg feed for 2 years. The control group consisted of 96 rats/sex and was fed untreated diet for 2 years. After 6 and 12 months six rats/sex were sacrificed and after 18 months 12 rats/sex were sacrificed (controls, respectively, 12 and 24 rats/sex). The only effects observed were reductions in body weight and food consumption in males and females at 1000 mg/kg. No dose-related effects were observed on mortality, clinical chemistry, haematology, clinical chemistry, organ weights, macroscopy or microscopy. No difference between control and treated groups were found in the sciatic nerves. There was no increase in compound-related tumours. The NOEL was 100 mg/kg feed, equivalent to 5 mg/kg bw per day (McAusland *et al.*, 1978).

2.2.3.3   Dogs

Groups of four beagle dogs/sex (5½-7 months of age) were fed diets containing 0, 3, 30, 300 or 1000 mg cypermethrin/kg food for 2 years. Groups of four dogs/sex were allocated to a satellite study and received a diet containing 0, 300 or 1000 mg/kg cypermethrin (data concerning this satellite group have only been included in the report to aid interpretation of the results of the main study). Due to severe signs of intoxication observed at the 1000 mg/kg level, the concentration was reduced to 750 mg/kg at week 4 of the study and, when signs of intoxication persisted during weeks 6-8, animals in the high-dose group were fed a control diet for 10 days to allow them to recover. Following the 10 days of control diet, the dogs were fed cypermethrin at a concentration of 600 mg/kg food for the remainder of the study. No effects were observed on ophthalmoscopy, clinical chemistry, organ weights, macroscopy or microscopy. No abnormalities were found in the sciatic nerves, brain or spinal cord. One male dog in the satellite group fed 1000 mg/kg convulsed and died. Signs of intoxication appeared within 24 hours of the initiation of dosing and consisted of licking and chewing of the paws, a stiff high stepping gait, whole body tremors, head shaking, incoordination, ataxia and, in some cases, convulsions. These signs were observed at 1000 mg/kg and to a lesser extent at 750 mg/kg, but not at 600 mg/kg. The body weights of male dogs in the highest dose group were significantly lower than the controls, probably due to initial reduction in food consumption observed at 1000 mg/kg. In the satellite group no effect on body weight gain was seen. Apart from some occasionally decreased sodium levels in the males given the highest dose, no consistent haematological changes were seen. The NOEL was 300 mg/kg feed, equivalent to 7.5 mg/kg bw per day (Buckwell, 1981).

2.2.4   Reproductive toxicity studies

Groups of 30 rats/sex (Wistar) received diets containing 0, 10, 100 and 500 mg cypermethrin/kg feed for 5 weeks prior to mating and then throughout pregnancy and lactation for three successive generations. Two litters were bred per generation. The first litters were discarded at weaning. Males and females from the second litter were randomly selected to breed the next generation. A significant reduction in body weight gain was seen in the male and female parent rats receiving 500 mg/kg in all three generations. This was correlated with a reduction in food consumption. Litter size was reduced at 500 mg/kg in the F_{1a} litter at birth and after 7 and 21 days. Litter weights were reduced at 500 mg/kg in the F_{1a} litters on days 7, 14 and 21 of lactation. No other effects on fertility or reproduction parameters were found. The NOEL for maternal and reproduction toxicity was 100 mg/kg feed, equivalent to 5 mg/kg bw per day (Hend *et al.*, 1978; Fish, 1979; Thorpe, 1985).

2.2.5   Special studies on embryotoxicity and teratogenicity

2.2.5.1   Rats

a)   Cypermethrin

Groups of 25 pregnant female rats (Sprague-Dawley) received by gavage 0, 17.5, 35 or 70 mg cypermethrin/kg bw per day in corn oil during days 6 to 15 of gestation. The females were sacrificed on day 21 of gestation for examination of their uterine contents. One female at 70 mg/kg bw per day was found dead and one female at 70 mg/kg bw per day was killed for ethical reasons following severe convulsions. Eleven out of 25 females at the 70 mg/kg bw per day group showed neurological disturbances (ataxia, convulsions, hypersensitivity to noise). A dose-related reduction in body weight gain was observed in the groups given 35 and 70 mg/kg bw per day. There were no indications of any embryotoxic or teratogenic effects. The NOEL for maternal toxicity was 17.5 mg/kg bw per day and the NOEL for embryotoxicity was 70 mg/kg bw per day (Tesh *et al.*, 1978).

b)   α-Cypermethrin

In a range-finding study, five pregnant female Sprague-Dawley rats received daily, by gavage, 0, 3, 9, 15 or 18 mg α-cypermethrin/kg bw in corn oil during days 6-15 of gestation. Maternal body weights, food consumption and clinical observations were recorded. On day 20 of gestation the females were killed and the fetuses were weighed, sexed and externally examined. Four dams at 18 mg/kg bw per day and one dam at 15 mg/kg bw per day showed hindlimb splay and unsteady gait during dosing. Mean body weight gain was reduced in a dose-related manner at 9, 15 and 18 mg/kg bw per day and at 15 and 18 mg/kg bw per day food consumption was reduced. No other treatment-related abnormalities were observed (Irvine & Twomey, 1994).

In another study, groups of 24 pregnant female Sprague-Dawley rats received by gavage 0, 3, 9 or 18 mg α-cypermethrin (purity 95.6%)/kg bw per day in corn oil during days 6-15 of gestation. Following marked clinical signs of toxicity the dose level of 18 mg/kg bw per day was lowered to 15 mg/kg bw per day on day 10 of gestation. Clinical signs, body weights and food consumption were recorded. On day 20 of gestation the females were killed and necropsied. The fetuses were weighed, sexed and examined for external, visceral and skeletal abnormalities.

Females at 18 mg/kg bw per day showed unsteady gait, piloerection, limb splay and hypersensitivity to sound. After reduction of the dose level the signs were similar but less marked. After treatment with 18/15 mg/kg bw per day a lowered body weight gain and food consumption was seen. At 9 mg/kg bw per day a slight body weight reduction was seen. Mean fetal weights were slightly reduced at 18/15 mg/kg bw per day. There was no indication for teratogenicity. The NOEL for maternal and fetal toxicity was 9 mg/kg bw per day (Irvine, 1994c).

2.2.5.2 Rabbits

a) Cypermethrin

In a range-finding study, groups of four female pregnant rabbits (NZW) received during days 6 to 18 of gestation by gavage 0, 25, 50, 100 or 120 mg cypermethrin/kg bw per day in corn oil. The dams were sacrificed on day 29 of gestation. No adverse effects were seen in the mothers and fetuses (Tesh *et al.*, 1984a).

Groups of 16 pregnant NZW rabbits received by gavage 0, 20, 50 or 120 mg cypermethrin/kg bw per day in corn oil during days 6 to 18 of gestation. The dams were killed on day 29 of gestation. One control female, three females receiving 20 mg/kg bw per day and two females in each of the groups receiving 50 and 120 mg/kg bw per day were killed for ethical reasons. Necropsy revealed evidence of respiratory tract infection and/or gastrointestinal tract infection not related to the substance. Two females at 20 mg/kg bw per day and two females at 120 mg/kg bw per day aborted during the post-treatment phase of the investigation. The number of implantations, live young and resorptions, pre- and post-implantation losses, and fetal and placental weights were unaffected by treatment. There was no indication for embryotoxicity or teratogenicity. The NOEL for embryo-toxicity was 120 mg/kg bw per day (Tesh *et al.*, 1984b, 1988).

Groups of pregnant rabbits (20 rabbits/group, 30 rabbits were used as an additional control group) were administered cypermethrin dissolved in corn oil at dose levels of 0, 3, 10 or 30 mg/kg bw per day orally from days 6 to 18 of gestation. On day 28 of gestation the rabbits were killed and examination was made of live fetuses, dead fetuses, resorption sites and corpora lutea. Live fetuses were maintained for 24 hours to assess viability. Fetuses were also examined for gross somatic and skeletal deformities. There was no significant mortality or difference in weight gain during the period of gestation. There were no significant differences between control and test groups with respect to pregnancy, fetal death and survival. Although a wide range of skeletal and visceral abnormalities was found in the course of the study, there were no differences between control and test groups with respect to abnormalities. It was concluded that oral dosing up to 30 mg/kg bw during the major period of organogenesis resulted in no teratogenic effects in offspring (FAO, 1980).

b)   α-Cypermethrin

In a range finding study groups of five mated female NZW rabbits received by gavage 0, 5, 15, 25 or 30 mg α-cypermethrin/kg bw per day as solutions in corn oil, during days 7-19 of gestation. On day 28 of pregnancy the females were killed and a necropsy was performed. The fetuses were weighed, sexed and externally examined. One female each at 15 and 25 mg/kg bw per day was killed prematurely. At 25 and 30 mg/kg bw per day marked reductions in body weight and food consumption were seen. There was no indication for either embryotoxicity or teratogenicity (Irvine, 1994a).

In another study, groups of 16 pregnant NZW rabbits received by gavage 0, 3, 15 or 30 mg α-cypermethrin/kg bw per day in corn oil during days 7-19 of gestation. Maternal clinical signs, body weight and food consumption were recorded. The females were killed on day 28 of pregnancy. The uterus was weighed and the numbers of corpora lutea, implantations and live fetuses were counted. The fetuses were weighed, sexed and examined for external, visceral and skeletal abnormalities.

Two control females, three at 15 mg/kg bw per day and two at 30 mg/kg bw per day were killed, because of severe weight loss and low food consumption. One female at 15 mg/kg bw aborted on day 28. In all groups, including controls, there was a similar mean body weight loss after the onset of dosing, which continued until day 11. At 30 mg/kg bw per day there was a further reduction in mean body weight gain towards the end of the dosing period. Food consumption reflected the changes in mean body weight gain. The NOEL for maternal toxicity was 3 mg/kg bw per day and the NOEL for embryotoxicity was 30 mg/kg bw. There was no indication for teratogenicity (Irvine, 1994b).

2.2.6   Special studies on genotoxicity

Results of genotoxicity studies carried out cypermethrin and α-cypermethrin are summarized in Tables 3 and 4.

Table 3. Results of mutagenicity assays on cypermethrin

Test system	Test object	Concentration	Purity (%)	Results	References
In vitro					
Gene muta-tions assay	*S. typhimurium* TA1525, TA100, TA1538, TA98, TA1537	0.2-2000 μg/plate[1]	93.5	negative[2]	Brooks, 1980 Dean, 1981
Gene muta-tions assay	*E.coli* WP2 or WP2 uvrA	0.2-2000 μg/plate[1]	93.5	negative[2]	Dean, 1981
Mitotic gene conversions assay	*Saccharomyces cerevisiae* JD1	0.01-5.0 mg/litre[1]	93.5	negative	Dean, 1981
Host-medi-ated assay in mice	*S. cerevisiae* JD1	25 and 50 mg/kg bw	93.5	negative	Brooks, 1980
Cell trans-formation assay	BHK 21/C113 cells	31.25-250 μg/ml[1]	93.5	negative	Dean, 1981
Chromoso-mal aberr-ations assay	RL$_4$ liver cells	7.5-30 μg/ml	93.5	negative	Dean, 1981
In vivo					
Chromosome aberration assay in bone marrow	Chinese hamster	2x oral dose of 20 or 40 mg/kg bw (2 successive days)	?	negative	Dean, 1977
Dominant lethal assay	mouse	single oral dose of 6.25, 12,5 or 25 mg/kg bw or 5 daily doses of 2.5 or 5 mg/kg bw per day	?	negative	Dean *et al.*, 1977
Host-medi-ated assay	mouse/ *S. cerevisiae*	orally 0, 25, 50 mg/kg	?	negative	JMPR, 1979

Table 3 (contd).

Test system	Test object	Concentration	Purity (%)	Results	References
DNA damage assay	CD rats	♂ 100 mg/kg, ♀ 150 mg/kg; 1, 4, 16 hour exposure	?	negative	Creedy & Wooder, 1977
	Wistar rats	♂ 300 mg/kg; 1, 4, 16 hour exposure ♀ 450 mg/kg, 1 hour exposure ♀ 337.5 mg/kg, 4 and 16 hour exposure		negative	Creedy & Wooder, 1977

[1] with and without metabolic activation
[2] at 200 and 2000 µg/plate, formation of visible droplets in the top agar was seen

2.2.7   Special studies on neurotoxicity

2.2.7.1   Rats

a)   Cypermethrin

Groups of 6 or 12 rats/sex were administered single oral doses of 100, 200 or 400 mg cypermethrin/kg bw (purity 97%) as a 5% dispersion in corn oil. The observation period was 9 days. The rats were then killed and examined histologically. All rats showed signs of intoxication. At 400 mg/kg bw, within 4 hours of dosing, rats developed signs of intoxication, including coarse tremors, spasmodic movements of the body and tail and bleeding from the nose. Tip-toe walking was also seen in some rats. All animals, except one, were killed, owing to the severity of the signs. Histological examination revealed swelling of the myelin sheaths and breaks of some of the axons of the sciatic nerves. At 200 mg/kg bw similar effects were observed; eight rats of each sex died or were killed within 48 h of dosing. The remaining four rats survived the observation period. At 100 mg/kg bw all animals survived the 9 days. One female out of 12 showed minimal lesions in the sciatic nerve in this group. A NOEL could not be determined (Carter & Butterworth, 1976).

In a neuromuscular dysfunction test, Wistar rats (10/sex) were treated by gavage with 0, 25, 50, 100, 150 or 200 mg cypermethrin/kg bw per day in DMSO for 7 consecutive days. The rats were killed 3-4 weeks after the start of dosing, and right and left sciatic/posterior tibial nerves were analysed. Neuromuscular function was assessed by means of the

Table 4. Results of mutagenicity assays on α-cypermethrin

Test system	Test object	Concentration	Purity (%)	Results	References
In vitro					
Gene mutations assay[1]	S. typhimurium TA98, TA100, TA1535, TA1537 TA1538 E. coli WP2 uvrA	31.5-5000 μg/ml	95.6	negative (no toxicity)	Brooks & Wiggins, 1992
Gene mutations assay[1]	S. cerevisiae XV185-14C	31.25-4000 μg/ml	95.8	negative	Brooks, 1984
Gene mutations[1] assay	L5178Y mouse lymphoma cells	3.3-50 μg/ml	95.4	negative	Vanderwaart, 1994
Chromosomal aberrations assay[1]	human peripheral lymphocytes	-act: 93.75-1000 μg/ml +act: 125-1000 μg/ml	95.6	negative precipitation was seen	Brooks & Wiggins, 1993
In vivo					
Chromosomal aberrations	rat femoral bone marrow	single oral dose 2-8 mg/kg[3]	95.8	negative	Clare & Wiggins, 1984
Micronucleus assay	mouse	single oral dose 1-10 mg/kg bw	95.4	negative	Vanderwaart, 1995
Alkaline elution analysis assay[2]	rat	single oral dose 40 mg/kg, 6 hours exposure	96.5	negative	Wooder, 1981

[1] With and without metabolic activation
[2] The effect of α-cypermethrin on the integrity of rat liver DNA was investigated by this method
[3] Initially doses of 10, 20 or 40 mg/kg were used, but the animals exhibited severe signs of toxicity and the experiment was terminated at these doses. Surviving females were evaluated for chromosomal damage and none was observed

inclined plane test and peripheral nerve damage by reference to β-glucuronidase and β-galactosidase activity increases in nerve tissue homogenates.

At 200 mg/kg bw per day 50% of the males and 62.5% of the females died, at 150 mg/kg bw per day two females and one male died and at 100 mg/kg bw per day one female died but no males. No mortalities occurred in the other groups. A dose-related (in severity and duration) increase was seen in clinical signs at doses ≥ 100 mg/kg bw. These signs included salivation, ataxia, splayed hind limb gait, hyperexcitability to auditory stimuli, tremor and choreoathetosis. At the same dose levels body weight gain was also reduced.

A dose-related transient functional impairment was found in rats treated with cypermethrin in the inclined plane test. This effect was maximal at the end of the 7-day subacute dosing regimen. At doses which caused mortality, significant increases in β-glucuronidase and β-galactosidase were found 3-4 weeks after the start of dosing in the distal portion of the sciatic/posterior tibial nerves. There was no direct correlation between the time course of the neuromuscular function and the neurobiochemical changes. The NOEL was 50 mg/kg bw per day (Rose & Dewar, 1983).

The effects of some pyrethroids, including cypermethrin, on amplitude and pre-pulse inhibition of the acoustic startle reflex were studied in male Wistar rats. The pyrethroids were suspended in corn oil. Cypermethrin was given orally at dose levels of 0, 0.5, 1 or 2 mg/kg bw to 12 males. Each animal received all doses of cypermethrin. The intersession interval was one week. Animal behaviour was observed by the experimenter before and after the test session for a period of 10 minutes and the effects of the pyrethroids on overt behaviour were measured by scoring the presence of pawing and salivation, burrowing, hyperactivity, hyperreactivity to an external stimulus, fine tremors of low intensity and fluid loss. Neither cypermethrin nor the other pyrethroids tested affected the amplitude or the latency of the startle reflex (Hijzen et al., 1988).

b)   α-Cypermethrin

The acute neurotoxicity of α-cypermethrin was studied in Crl:CD:BR rats in two separate acute studies, each using four groups of ten animals/sex or five rats per sex (additional study). The groups received a single dose of 0, 4, 20 or 40 mg α-cypermethrin/kg bw in corn oil. During the 14-day observation period, clinical signs and body weight were analysed. In the main study a detailed clinical assessment for neurotoxicological effects was performed. This included a functional observational battery (FOB) and measurements of fore and hind limb grip strength, hindlimb landing foot splay and motor activity. In each study five rats/sex were killed on day 15 and brain, eyes, muscle, nerves, spinal cord and spinal ganglia were analysed.

One male rat in each of the 20 and 40 mg/kg bw groups of the additional study was found dead on the day after dosing. Clinical signs were seen in male rats dosed with 20 and 40 mg/kg bw. The signs (similar in both studies) developed between 3 to 8 hours after dosing and resolved by three days after dosing. The signs included abnormal/splayed gait, thrashing, prostration, vocalization, piloerection, hunched posture, unkempt appearance, soiled/stained body areas and diarrhoea. The signs in females were similar but lower in frequency. In addition to these signs there were also isolated cases of twitching, tremors, abasia, hypersensitivity, pale eyes, soft faeces and thinning of the fur.

During FOB conducted 5 hours after dosing, gait abnormalities and clinical signs of increased reactivity were seen in most male rats dosed with 20 and 40 mg/kg bw. In females the signs were less frequent. In the 20 and 40 mg/kg bw groups there was a increase in very slight to slight sporadic fibre degeneration in the sciatic nerve. The changes were more frequent in the proximal than in the distal part of the nerve. The NOEL was 4 mg/kg bw (Fokkema, 1994b).

c)   Cypermethrin and α-cypermethrin

This experiment was performed in two phases. The first phase was conducted to determine the time course for neurochemical changes in Wistar rats occurring in the sciatic/posterior tibial nerve (SPTN), trigeminal nerve and trigeminal ganglion following treatment with cypermethrin for 5 days/week for 4 weeks. Cypermethrin was administered at 150 mg/kg bw per day in DMSO (reduced to 100 mg/kg bw per day in arachis oil, after 10 doses, because of mortality). α-Cypermethrin was dosed at 37.5 mg/kg bw per day in DMSO (also reduced after 10 days to 25 mg/kg bw per day in arachis oil). Five animals per sex, treated with either cypermethrin and α-cypermethrin, were killed at 2, 3, 4, 5, 6, 8, 10 or 12 weeks and examined.

Dosing resulted in the death of 56% of the cypermethrin-treated animals and 21% of the α-cypermethrin-treated animals. The most frequent signs of intoxication included abnormal gait, ataxia, lethargy, chromodacryorrhoea, salivation and hypersensitivity to sensory stimuli. The β-glucuronidase and β-galactosidase activities in the SPTN were increased at 5, 6 and 8 weeks, when compared to controls. The increase was maximal after 5 weeks, and after 12 weeks was comparable to controls. No significant enzyme changes were found in the trigeminal ganglia and trigeminal nerve of treated animals.

Phase 2 was conducted to establish the dose level which did not cause peripheral nerve degeneration in the SPTN, trigeminal nerve and ganglia. Groups of 10 rats/sex were dosed with 37.5, 75 or 150 mg cypermethrin/kg bw per day in DMSO or 10, 20 or 40 mg α-cypermethrin/kg bw per day in DMSO, 5 days/week for 4 weeks. A control group of 10 animals was used.

Signs of intoxication similar to those reported in phase 1 were seen at the highest dose levels. A large increase in β-glucuronidase and β-galactosidase activities in the SPTN was seen at 150 mg/kg bw per day cypermethrin and 40 mg/kg bw per day α-cypermethrin. In the groups administered 75 mg/kg bw per day cypermethrin or 20 mg/kg bw per day α-cypermethrin a small increase in β-galactosidase was found in both the distal and proximal sections of the SPTN. The magnitude of the enzyme changes was similar to those of phase 1. Significant enzyme changes were also found in the trigeminal ganglia and to a lesser extent in the trigeminal nerve of the groups administered 75 or 150 mg/kg bw per day cypermethrin and 20 or 40 mg/kg bw per day α-cypermethrin. The NOELs were 37.5 mg/kg bw cypermethrin and 10 mg/kg bw α-cypermethrin (Rose, 1983).

2.2.7.2   Chickens

In a delayed neurotoxicity study, six adult domestic hens received 1000 mg cypermethrin/kg bw per day in DMSO for 5 days. After 3 weeks the dosing regime was repeated and a further three weeks later the birds were killed. A positive (tri-*ortho*-tolyl phosphate) and negative control (not dosed) group were used.

None of the cypermethrin-treated hens developed any signs of intoxication. Histological examination of the nervous system revealed no lesions. All birds receiving the positive control developed clinical signs of neurological damage within 15 days and became progressively more unsteady and ataxic thereafter. Histological examination of these animals showed lesions in the cerebellum, sciatic nerve and spinal cord, including axonal and myelin degeneration (Owen & Butterworth, 1977).

2.2.7.3   Hamsters

The 1979 JMPR (FAO, 1980) evaluated some neurotoxicity studies with cypermethrin in hamsters. At doses of ≥ 794 mg/kg bw, all treated hamsters showed clinical signs of poisoning, including tremors, abnormal irregular movements and an unusual gait. As in the cases of rats, axon and myelin degeneration was noted in all groups treated. The lesions included swelling and breaks in the axons and clumping of myelin.

Hamsters treated orally with a single dose of 40 mg/kg bw, followed by four doses of 20 mg/kg bw, developed weight loss and sometimes mortality. There was loss of fur and dermal ulceration. There was no effect in the mean slip angle experiment, and a marginal increase in β-galactosidase was observed in peripheral nerve.

Hamsters treated orally with doses of 5, 10 or 20 mg/kg bw per day for 5 days showed no mortality. Lower body weight gain was observed at 20 mg/kg bw per day. One female (out of 5) at the highest dose developed hyperexcitability. There was a significant deficit in the mean slip angle test, females showing an earlier dose-related deficit than noted in males. β-Galactosidase activity was increased at all dose levels 3 weeks after the onset of the experiment. This effect was significant at the two highest dose levels. In this experiment dermal irritation and fur loss were noted.

In another experiment hamsters were orally treated with 30 mg/kg bw per day for 5 days. There was no mortality and there were no differences in weight gain. There was some transient skin irritation accompanied by skin ulceration. One male (out of 16) had an unusual gait. There was a slight deficit in the inclined plane test which was noted in the early parts of the experiment. Increases in both β-glucuronidase and β-galactosidase were evident in peripheral nerve tissue.

2.2.8   Special studies on biochemistry and electrophysiology

In three independent experiments with Wistar rats the effects of varying doses of cypermethrin (purity 98%, 10% in DMSO) on the trigeminal ganglion and three sections of the maxillary branch of the trigeminal nerve (proximal, distal and endings) were determined. Increases in β-galactosidase activity in these tissues were taken as evidence of axonal degeneration.

The three studies involved repeated oral administration of cypermethrin at 150 mg/kg bw per day for 5 or 7 days and 0, 25, 50, 100 or 200 mg/kg bw per day for 5 or 7 days. Mortality occurred in animals receiving 100 mg/kg bw per day or more. A dose-related transient functional impairment, assessed by means of the inclined plane test was found in the first week. Significant increases in β-glucuronidase and β-galactosidase activity of the sciatic, tibial or trigeminal nerves only occurred with 5 or 7 doses of 150 or 200 mg/kg bw per day. Increased activity of the enzymes in the distal portion of nerves was found, but even in the most severely intoxicated animals the magnitude of this increase was less than that induced by the known neurotoxic agent methylmercury chloride (Dewar & Moffett, 1978).

Cypermethrin (1:1 cis:trans) was administered to male and female rats at dose levels ranging from 25 to 200 mg/kg bw per day for 5 consecutive days by oral intubation as a 10% w/v solution in DMSO. A dose-related functional deficit was observed when the mean slip angle test and the landing foot spread test were applied to the animals. The deficit was maximal from days 6 to 14 after the beginning of treatment, and complete functional recovery occurred within 4 weeks. Substantial variation in data from the landing foot spread test was noted. Data were inconsistent over the course of the study. β-Glucuronidase activity was increased in a dose-dependent fashion in both males and females. The results suggested that cypermethrin produced a primary axonal degeneration, readily measurable 28 days after treatment as an increase in β-glucuronidase activity and in deficits in specific behavioural-function testing of rats (FAO, 1980).

Electrophysiological studies were performed to determine whether acute or subacute intoxication with cypermethrin produced changes in the conduction velocity of slower fibres in peripheral nerves or alterations in the maximal motor conduction velocity. There was no evidence to suggest that cypermethrin, at doses that induced severe clinical signs of intoxication, including ataxia, had any effect on maximal motor conduction velocity or conduction velocity of the slower motor fibres in peripheral nerves. Doses used in the study ranged from a single dose of 200 mg/kg bw to 7 consecutive doses of 150 mg/kg bw followed by two doses of 400 mg/kg bw. At near-lethal doses there were no effects on conduction velocity even in the presence of clinical signs of acute intoxication and at dose levels where previous studies had shown functional degeneration. These electrophysiological findings are reflective of motor function, which would suggest that the physiological and functional deficits observed as a result of acute intoxication are primarily sensory in nature (FAO, 1980).

2.2.9   Special studies on sensitization

2.2.9.1   Cypermethrin

Two out of 20 guinea-pigs developed a positive reaction to cypermethrin in the Magnusson Kligman test, indicating that cypermethrin is not a sensitizer (Coombs *et al.*, 1976).

2.2.9.2   α-Cypermethrin

No positive reactions were obtained in a Magnusson-Kligman test performed with guinea-pigs (Gardner, 1993).

2.2.10   Special studies on skin and eye irritation

2.2.10.1   Cypermethrin

A single application of undiluted technical cypermethrin was moderately irritant to occluded rabbit skin (Coombs *et al.*, 1976).

A single application of undiluted technical cypermethrin to rabbit eyes produced a mild transient conjunctivitis and blepharism lasting 2 days (Coombs *et al.*, 1976).

2.2.10.2   α-Cypermethrin

Six NZW rabbits receiving a semi-occlusive topical application with 500 mg α-cypermethrin technical developed very slight erythema in two animals up to 72 hours after removal of the dressings. There were no other dermal reactions (Gardner, 1993).

Six NZW rabbits receiving an instillation of 0.1 ml (equivalent to 45 mg) α-cypermethrin technical developed slight conjunctival redness and ocular discharge up to 72 hours after treatment. No cornea or iris irritation was observed (Gardner, 1993).

2.3   Observations in humans

The symptoms and signs of acute poisoning resulting from pyrethroids are very similar. Apart from the irritative symptoms of the skin and respiratory tract (or digestive tract in ingestive poisoning), acute pyrethroid poisoning is clinically characterized by abnormalities of nervous excitability.

Occupationally exposed people had abnormal skin sensations described as burning, itching or tingling, which could be exacerbated by sweating or washing and readily disappeared after several hours. Systemic symptoms included dizziness, headache, nausea, anorexia and fatigue. Vomiting was more prominent in patients with ingestive poisoning than in occupational poisoning cases. Other symptoms such as chest tightness, paraesthesia, palpitation, blurred vision and increased sweating were less frequently seen. The more serious cases developed coarse muscular fasciculations in large muscles or extremities (Van den Bercken & Vijverberg, 1989; He et al., 1989).

Urine obtained from operators spraying cypermethrin in experimental trials was analysed for the presence of the chlorinated cyclopropane carboxylic acid metabolite. This metabolite was observed in the urine of exposed workers at levels up to 0.4 μg/ml (the limit of detection was estimated to be 0.05 μg/ml).

Cypermethrin sprayers were found to have residues on the exposed parts of their bodies. The rate of dermal exposure of the operators during spraying ranged from 1.5 to 46.1 mg/hour. There was a reasonable relationship between the total cypermethrin deposited dermally and the excretion in urine. The levels of the cyclopropane carboxylic acid metabolite in the 24-hour urine were between <0.05 and 0.32 mg (not specified). This, together with the finding of 0.6 mg (not specified) of this metabolite in 72-hour urine from one man, led to the estimation that approximately 3% of the total dermal dose was absorbed and rapidly excreted by the operators (FAO, 1980).

In a controlled experiment with sprayers, no abnormalities were found in clinical and neurological examinations, blood chemistry or peripheral nerve function tests (including the trigeminal nerve). In some electroneurophysiological tests (motor conduction velocity, slow fibre conduction velocity and cornea reflex), a significant change within the normal range appeared to exist for the group sprayers between pre- and post-exposure measurements. These changes probably reflect seasonal variations (FAO, 1982).

3.   COMMENTS

The Committee considered toxicological data on cypermethrin and α-cypermethrin, including the results of acute, short-term, and reproductive studies, and studies on pharmacokinetics and metabolism, genotoxicity, long-term toxicity/carcinogenicity and neurotoxicity. Results of effects in humans were also considered.

Cypermethrin is a mixture of four cis and four trans isomers. The cis isomers are more biologically active and more persistent than the trans isomers. α-Cypermethrin is a mixture of the two most active cis isomers. A typical cypermethrin sample contains 25% α-cypermethrin. Cypermethrin and α-cypermethrin are α-cyano or type II pyrethroids that cause neurotoxicity in mammals and insects. They affect nerve membrane sodium channels, causing a long-lasting prolongation of the normally transient increase in sodium permeability of the membrane during excitation. At high dose levels, these type II pyrethroids induce salivation and tremors that progress to characteristic clonic-tonic convulsions (choreoathetosis and salivation syndrome).

After oral administration, cypermethrin is readily absorbed, distributed and excreted in rats, chickens, sheep and cattle. Cypermethrin is primarily eliminated in urine and faeces in about equal proportions. Less than 1% is excreted in milk. When cypermethrin was applied dermally to sheep, 2.5% was eliminated in urine and faeces within 6 days and after an oral dose about 60% was eliminated within 2 days.

Studies in cattle indicated that absorption, distribution and excretion were comparable for cypermethrin and α-cypermethrin. The major metabolic route for both cypermethrin and its isomers, including α-cypermethrin, is cleavage of the ester bond followed by hydroxylation and conjugation of the cyclopropyl and phenoxybenzyl portions of the molecule. The data suggest that there is no isomeric interconversion during metabolism.

The acute oral toxicity of cypermethrin and α-cypermethrin is moderate to high. WHO has classified these substances as "moderately hazardous" (WHO, 1996). In rats and mice, the oral LD_{50} ranges from 82 to 4000 mg/kg bw for cypermethrin and from 35 to >5000 mg/kg bw for α-cypermethrin, depending on the vehicle used. At lethal or near lethal doses the signs are typical of type-II pyrethroids and include salivation, ataxia, gait abnormalities and convulsions.

Several oral short-term toxicity studies with cypermethrin were available. Cypermethrin has been tested in rats (5 weeks and 90 days) and dogs (5 weeks and 13 weeks) at dose levels ranging from 25 to 1600 mg/kg feed (equivalent to 1.25-80 mg/kg bw per day) and 5 to 1500 mg/kg feed (equivalent to 0.125-37.5 mg/kg bw per day), respectively. In these studies, the clinical signs included ataxia, abnormal gait, nervousness, and, particularly in dogs, inappetence, diarrhoea, vomiting and hyperaesthesia. In both rats and dogs, cypermethrin caused decreases in body weight gain, food intake, and a number of haematological parameters, increases in some organ weights and plasma urea levels, and, at lethal or near-lethal doses, effects on the nervous system. For cypermethrin the lowest NOEL in short-term studies was in a 90-day study with rats administered 25, 100, 400 or 1600 mg/kg feed (equivalent to 1.25-80 mg/kg bw per day). Male rats given 1600 mg/kg feed showed decreases in haemoglobin concentration, mean corpuscular volume and eosinophil numbers, and increases in prothrombin time, plasma urea levels and relative liver and kidney weights. The decrease in eosinophil numbers and increase in relative liver weight were also observed in males at 400 mg/kg feed. In female rats, reduced food intake and increased relative liver weight were noted in rats given 1600 mg/kg feed. The NOEL in this study was 100 mg/kg feed, equivalent to 5 mg/kg bw per day.

α-Cypermethrin was tested in oral short-term toxicity studies with mice (29 days and 13 weeks), rats (5 weeks, 6 weeks and 90 days) and dogs (13 weeks and 52 weeks) at dose levels ranging from 50 to 1600 mg/kg feed (equivalent to 7-240 mg/kg bw per day), 20 to 1200 mg/kg feed (equivalent to 1.25-60 mg/kg bw per day) and 30 to 270 mg/kg feed (equivalent to 0.75-6.75 mg/kg bw per day), respectively. In these studies, α-cypermethrin caused the same effects as described for cypermethrin in the short-term studies. The signs of toxicity included ataxia, abnormal gait, increased sensitivity to noise, hyperactivity, hunched posture and, as demonstrated histologically, axonal degeneration of the sciatic nerves. For α-cypermethrin the lowest NOEL was in a 52-week study with dogs. In this study a diet containing 60, 120 or 240 mg/kg feed (equivalent to 1.5 to 6 mg/kg bw per day) was administered. Dogs given 120 and 240 mg/kg feed showed skin reddening, ulceration and necrosis. The NOEL in this study was 1.5 mg/kg bw per day.

A three-generation reproductive toxicity study with cypermethrin was performed in rats at dose levels of 10, 100 or 500 mg/kg feed (equivalent to 0.5-25 mg/kg bw per day). At the highest dose a reduction in body weight gain and food consumption and a concomitant reduction in litter size and weight were seen in the F_{1a} progeny only. No other effects on fertility or reproduction parameters were observed. The NOEL was 100 mg/kg feed, equivalent to 5 mg/kg bw per day.

Cypermethrin did not cause embryotoxicity or teratogenicity in rats at doses up to 70 mg/kg bw per day or in rabbits at doses up to 120 mg/kg bw per day. α-Cypermethrin did not cause embryotoxicity or teratogenicity in rats at doses up to 9 mg/kg bw per day or in rabbits up to 30 mg/kg bw per day. The NOELs for maternal toxicity in rats were 17.5 and 9 mg/kg bw per day for cypermethrin and α-cypermethrin, respectively, while the NOELs for maternal toxicity in rabbits were 30 and 3 mg/kg bw per day for cypermethrin and α-cypermethrin, respectively.

Cypermethrin and α-cypermethrin have been tested in a wide variety of *in vitro* and *in vivo* genotoxicity studies. All of the results were negative.

Two long-term toxicity/carcinogenicity studies with mice and rats were available on cypermethrin. Mice received a diet containing 100, 400 or 1600 mg/kg feed (equal to 14-228 mg/kg bw per day) for 101 weeks. At 1600 mg/kg feed reduced body weight gain, changes in haematological parameters and increased liver weight were observed. No effects were observed at 400 mg/kg feed, equal to 57 mg/kg bw per day. In a study in which rats received diets containing 1, 10, 100 or 1000 mg/kg feed (equivalent to 0.05-50 mg/kg bw per day) for 2 years, the only effects observed were reductions in body weight and food consumption at 1000 mg/kg feed. The NOEL was 100 mg/kg feed, equivalent to 5 mg/kg bw per day.

In a two-year toxicity study, dogs received diets containing 3, 30, 300 or 1000 mg cypermethrin/kg feed (equivalent to 0.075-25 mg/kg bw per day). The dose level of 1000 mg/kg feed was reduced to 600 mg/kg feed owing to severe intoxication. At 300 mg/kg feed, equivalent to 7.5 mg/kg bw per day, no effects were seen. The Committee concluded that cypermethrin was not carcinogenic in these studies.

Long-term toxicity/carcinogenicity or reproductive toxicity studies were not available on α-cypermethrin. The Committee noted the absence of reproductive toxicity and carcinogenicity associated with administration of cypermethrin, which contains 25% α-cypermethrin. The Committee also noted the absence of genotoxicity for either cypermethrin or α-cypermethrin, the absence of carcinogenicity associated with compounds of similar structure and the similar metabolism and disposition of the two compounds. In view of the foregoing, the Committee concluded that it was unnecessary to request the results of long-term toxicity/carcinogenicity or reproductive toxicity studies on α-cypermethrin.

Several studies on the neurotoxicity of cypermethrin and α-cypermethrin in rats were available. In these studies high oral doses of cypermethrin and α-cypermethrin caused clinical signs that included coarse tremor and spasmodic body and tail movements. Evidence of axonal damage in the sciatic/posterior tibial nerves and the trigeminal nerve and ganglion was indicated by significant increases in β-glucuronidase and β-galactosidase in nerve tissue homogenates, in addition to abnormal neuromuscular function tests. In the inclined plane test, cypermethrin (in

DMSO) caused transient functional impairment. The lowest NOEL for neurotoxicity was 37.5 mg/kg bw per day for cypermethrin (in DMSO) and 4 mg/kg bw per day for α-cypermethrin (in corn oil), indicating that the toxicity may be influenced by the vehicle used.

Humans occupationally exposed to cypermethrin developed skin sensation as a first reaction, followed by systemic effects such as dizziness, headache, nausea, paraesthesia and increased sweating. In more serious cases, muscular fasciculations developed in large muscles or in the extremities. In experiments with operators spraying cypermethrin, no clinical nervous system abnormalities were observed. However, exposure levels were not measured.

4.   EVALUATION

The Committee established an ADI of 0-50 µg/kg bw for cypermethrin on the basis of the NOEL of 5 mg/kg bw per day in 90-day, 2-year and reproductive toxicity studies in rats and the application of a safety factor of 100. The Committee established an ADI of 0-20 µg/kg bw for α-cypermethrin on the basis of the NOEL of 1.5 mg/kg bw per day in a 52-week study in dogs and the application of a safety factor of 100.

5.   REFERENCES

Amyes, S.J., Holmes, P., Irving, E.M., Green, C.F., Virgo, D.M., & Sparrow, S. (1994). Alpha-cypermethrin: preliminary toxicity study by dietary administration to CD-1 mice for 13 weeks. Unpublished report No. 92/SHL009/0849 from Pharmaco-LSR Ltd, Suffolk, England. Submitted to the WHO by Cyanamid, Wayne, NJ. USA.

Brooks, T.M. (1980). Toxicity studies with agricultural chemicals: Mutagenicity studies with RIPCORD in microorganisms *in vitro* and in the host-mediated assay (Experiment Nos 1846 and 1847). Unpublished report TLGR.80.059 from Shell Toxicology Laboratory (Tunstall). Submitted to the WHO by Cyanamid, Wayne, NJ, USA.

Brooks, T.M. (1984). Genotoxicity studies with Fastac: the induction of gene mutation in the yeast *Saccharomyces cerevisiae*. Unpublished report No. SBGR.84.117 from Shell Research Limited, Sittingbourne Research Centre. Submitted to the WHO by Cyanamid, Wayne, NJ, USA.

Brooks, T.M. & Wiggins, D.E. (1992). FASTAC TM: Bacterial mutagenicity studies. Unpublished report No. SBTR.92.022 from Shell Research Limited, Sittingbourne Research Centre. Submitted to the WHO by Cyanamid, Wayne, NJ, USA.

Brooks, T.M. & Wiggins, D.E. (1993). FASTAC TM: *In vitro* chromosome studies using cultured human lymphocytes. Unpublished report No. SBTR.93.007 from Shell Research Limited, Sittingbourne Research Centre. Submitted to the WHO by Cyanamid, Wayne, NJ, USA.

Buckwell, A.C. (1981). A 2-year feeding study in dogs on WL 43467 (Experiment No. 1412). Unpublished report No. SBGR.81.126 (plus two addenda) from Shell Toxicology Laboratory (Tunstall). Submitted to the WHO by Cyanamid, Wayne, NJ, USA.

Buckwell, A.C. & Butterworth, S. (1977). Toxicology studies on the pyrethroid insecticide WL 43467: a 13-week feeding study in dogs (Experiment No. 1112). Unpublished report No. TLGR 0127.77 (including addendum and corrigendum) from Shell Toxicology Laboratory (Tunstall). Submitted to the WHO by Cyanamid, Wayne, NJ, USA.

Carter, B.I. & Butterworth, S.T.G. (1976). Toxicity of insecticides: The acute oral toxicity and neuropathological effects of WL 43467 to rats. Unpublished report No. TLGR.0055.76 from Shell Research, Sittingbourne Research Centre. Submitted to the WHO by Cyanamid, Wayne, NJ, USA.

Clare, M.G. & Wiggins, D.E. (1984). Genotoxicity studies with fastac: *in vivo* cytogenetic test using rat bone marrow. Unpublished report No. SBGR.84.120 from Shell Research Limited, Sittingbourne Research Centre. Submitted to the WHO by Cyanamid, Wayne, NJ, USA.

Clark, D.G. (1982). WL85871: A 90-day feeding study in rats. Unpublished report No. SBGR.81.293 from Shell Research Limited, Sittingbourne Research Centre. Submitted to the WHO by Cyanamid, Wayne, NJ, USA.

Coombs, A.D., Carter, B.I., Hend, R.W., Butterworth, S.G., & Buckwell, A.C. (1976) Toxicity studies on the insecticide WL 43467: Summary of results of preliminary experiments. Unpublished report No. TLGR.0104.76 from Shell Toxicology Laboratory (Tunstall). Submitted to the WHO by Cyanamid, Wayne, NJ, USA.

Crawford, M.J. (1978). The excretion and residues of radioactivity in cows treated orally with ^{14}C-labelled WL 43467. Unpublished report No. TLGR.0029.78 from Shell Toxicology Laboratory (Tunstall). Submitted to the WHO by Cyanamid, Wayne, NJ, USA.

Crawford, M.J. & Hutson, D.H. (1977). The elimination and retention of WL 43467 when administered dermally or orally to sheep. Unpublished report No. TLGR.0098.77 from Shell Toxicology Laboratory (Tunstall). Submitted to the WHO by Cyanamid, Wayne, NJ, USA.

Crawford, M.J. & Hutson, D.H. (1978). The elimination of residues from the fat of rats following the oral administration of [^{14}C-benzyl]WL 43381 (cis-WL 43467). Unpublished report No. TLGR.0078.78 from Shell Toxicology Laboratory (Tunstall). Submitted to the WHO by Cyanamid, Wayne, NJ, USA.

Creedy, C.L. & Wooder, M.F. (1977) Studies on the effect of WL 43467 upon the integrity of rat liver cell DNA *in vivo*. Unpublished report No. TLGR.0043.77 from Shell Toxicology Laboratory (Tunstall). Submitted to the WHO by Cyanamid, Wayne, NJ, USA.

Croucher, A., Stoydin, G., Cheeseman, M.E., Buckwell, A.C., & Gellatly, J.B.M. (1980). The metabolic fate of cypermethrin in the cow: elimination and residues derived from ^{14}C-benzyl label. Unpublished report No. TLGR.80.121 from Shell Toxicology Laboratory (Tunstall). Submitted to the WHO by Cyanamid, Wayne, NJ, USA.

Dean, B.J. (1977). Toxicity studies with WL 43467: Chromosome studies on bone marrow cells of Chinese hamsters after two daily oral doses of WL 43467. Unpublished report No. TLGR.0136.77 from Shell Toxicology Laboratory (Tunstall). Submitted to the WHO by Cyanamid, Wayne, NJ, USA.

Dean, B.J. (1981). Toxicity studies with agricultural chemicals: *In vitro* genotoxicity studies with RIPCORD (Experiment No.AIR-1820). Unpublished report No. TLGR.80.116 from Shell Toxicology Laboratory (Tunstall). Submitted to the WHO by Cyanamid, Wayne, NJ, USA.

Dean, I. & Jackson, F. (1995). WL85871: 52-week oral (dietary) toxicity study in dogs (IRI Project No. 652238). Unpublished report No. IRI/11110 from Inveresk Research International, Scotland. Submitted to the WHO by Cyanamid, Wayne, NJ, USA.

Dean, B.J., Van der Pauw, C.L., & Butterworth S.T.G. (1977). Toxicity studies with WL 43467: Dominant lethal assay in male mice after single oral doses of WL 43467. Unpublished report No. TLGR.0042.77 from Shell Toxicology Laboratory (Tunstall). Submitted to the WHO by Cyanamid, Wayne, NJ, USA.

Dewar, A.J. & Moffett, B.J. (1978). Toxicity studies on the insecticide WL 43467: Biochemical studies on the effect of WL 43467 on the rat trigeminal nerve and ganglion. Unpublished report No. TLGR.0162.77 from Shell Toxicology Laboratory (Tunstall). Submitted to the WHO by Cyanamid, Wayne, NJ, USA.

Dunsire, J.P. & Gifford, L.J. (1993). The distribution of [benzyl-^{14}C]WL85871 (FASTAC) in the lactating cow (nature of residue study to EPA guidelines - Live phase) (IRI Project No. 153667). Unpublished report No. IRI/9879 from Inveresk Research International, Scotland. Submitted to the WHO by Cyanamid, Wayne, NJ, USA.

FAO (1980). Pesticide residues in food-1979. Report of the Joint Meeting of the FAO Panel of Experts on Pesticide Residues in Food and the Environment and the WHO Expert Group on Pesticide Residues. Rome, Food and Agriculture Organization of the United Nations (FAO Plant Production and Protection Paper, No. 20).

FAO (1982). Pesticide residues in food-1981. Report of the Joint Meeting of the FAO Panel of Experts on Pesticide Residues in Food and the Environment and the WHO Expert Group on Pesticide Residues. Rome, Food and Agriculture Organization of the United Nations (FAO Plant Production and Protection Paper, No. 37).

Fish, A. (1979). Corrigendum I in Group Research Report No. TLGR.0188.78. Unpublished data from Shell Toxicology Laboratory (Tunstall). Submitted to the WHO by Cyanamid, Wayne, NJ, USA.

Fokkema, G.N. (1994a). WL85871 (FASTAC): A 6-week range finding feeding study in the rat. Unpublished report No. SBTR.93.002 from Shell Research Limited, Sittingbourne Research Centre. Submitted to the WHO by Cyanamid, Wayne, NJ, USA.

Fokkema, G.N. (1994b). WL85871 (FASTAC): An acute oral (gavage) neurotoxicity study in the rat. Unpublished report No. SBTR.92.027 from Shell Research Limited, Sittingbourne Research Centre. Submitted to the WHO by Cyanamid, Wayne, NJ, USA.

Gardner, J.R. (1993). FASTAC technical: Acute oral and dermal toxicity in rat, skin and eye irritancy in rabbit and skin sensitisation potential in guinea pig. Unpublished report No. SBTR.92.033 from Shell Research Limited, Sittingbourne Research Centre. Submitted to the WHO by Cyanamid, Wayne, NJ, USA.

Green, C.F. (1993). Alphacypermethrin: Preliminary toxicity study by dietary administration to CD-1 mice for four weeks. Unpublished report No. LSR 92/0346 from Life Science Research Limited. Submitted to the WHO by Cyanamid, Wayne, NJ, USA.

Greenough, R.J. & Goburdhun, R. (1984). WL 85871: Oral (dietary) maximum tolerated dose study in dogs (IRI Project No. 631087).

Unpublished report No. IRI/3107 from Inveresk Research International, Scotland. Submitted to the WHO by Cyanamid, Wayne, NJ, USA.

Greenough, R.J., Cockrill, J.B., & Goburdhun, R. (1984). WL 85871: 13-week oral (dietary) toxicity study in dogs (IRI Project No. 631092). Unpublished report No. IRI/3197 from Inveresk Research International, Scotland. Submitted to the WHO by Cyanamid, Wayne, NJ, USA.

He, F., Wang, S., Liu, L., Chen, S., Zhang, Z. & Sun, J. (1989). Clinical manifestations and diagnosis of acute pyrethroid poisoning. *Arch. Toxicol.*, **63**, 54-58.

Hend, R.W., Hendy, R. & Fleming, D.J. (1978). Toxicity studies on the insecticide WL 43367: a three generation reproduction study in rats. Unpublished report No. TLGR.0188.78 from Shell Toxicology Laboratory (Tunstall). Submitted to the WHO by Cyanamid, Wayne, NJ, USA.

Henderson, C. & Parkinson, G.R. (1981). Cypermethrin technical: subacute dermal toxicity study in rabbits (CTL Study No. LB0019). Unpublished report No. CTL/P/588 from ICI Central Toxicology Laboratory. Submitted to the WHO by Cyanamid, Wayne, NJ, USA.

Hijzen, T.H., de Beun, R., & Slanger, J.L. (1988). Effects of pyrethroids on the acoustic startle reflex in the rat. *Toxicology*, **49**, 271-276.

Hutson, D.H. & Stoydin, G. (1976). The excretion of radioactivity from cows fed with radioactively labelled WL 43467. Unpublished report No. TLGR.0075.76 from Shell Toxicology Laboratory (Tunstall). Submitted to the WHO by Cyanamid, Wayne, NJ, USA.

Hutson, D.H. & Stoydin, G. (1987). Excretion and residues of the pyrethroid insecticide cypermethrin in laying hens. *Pestic. Sci.*, **18**, 157-168.

IPCS (1996) The WHO recommended classification of pesticides by hazard and guidelines to classification 1996-1997. World Health Organization, Geneva, p. 23 (Document WHO/PCS/96.3).

Irvine, L.F.H (1994a). Oral (gavage) rabbit developmental toxicity dose ranging study. Unpublished report No. SLN/3/92 from Toxicol Laboratories Limited, Ledbury, England. Submitted to the WHO by Cyanamid, Wayne, NJ, USA.

Irvine, L.F.H. (1994b). Oral (gavage) rabbit developmental toxicity (teratogenicity) study. Unpublished report No. SLN/4/93 from Toxicol Laboratories Limited, Ledbury, England. Submitted to the WHO by Cyanamid, Wayne, NJ, USA.

Irvine, L.H.F. (1994c). Alphacypermethrin. Oral (gavage) rat developmental toxicity (teratogenicity) study. Unpublished report No. SLN/2/92 from Toxicol Laboratories Limited, Ledbury, England. Submitted to the WHO by Cyanamid, Wayne, NJ, USA.

Irvine, L.F.H & Twomey, K. (1994). Alphacypermethrin. Oral (gavage) rat developmental study. Unpublished report No. SLN/1/92 from Toxicol Laboratories Limited, Ledbury, England. Submitted to the WHO by Cyanamid, Wayne, NJ, USA.

Jackson, G.C. (1993). Alphacypermethrin: Acute inhalation toxicity in rats, 4-hour exposure. Unpublished report No. SLL 266/930770 from Huntingdon Research Centre, England. Submitted to the WHO by Cyanamid, Wayne, NJ, USA.

Lindsay, S., Banham, P.B., Chart, I.S., Chalmers, D.T., Godley, M.J., & Taylor, K. (1982). Cypermethrin: Lifetime feeding study in mice (CTL Study No. PM0366). Unpublished report No. CTL/P/687 (including 1 supplement) from ICI Central Toxicology Laboratory. Submitted to the WHO by Cyanamid, Wayne, NJ, USA.

McAusland, H.E., Butterworth, S.T.G., & Hunt, P.F. (1978). Toxicity studies on the insecticide WL 43467: A two-year feeding study in rats. Unpublished report No. TLGR.0189.78 (including 4 corrigenda/addenda) from Shell Toxicology Laboratory (Tunstall). Submitted to the WHO by Cyanamid, Wayne, NJ, USA.

Morrison, B.J. & Richardson, K.A. (1994). WL85871 (alphacypermethrin, FASTAC): The metabolism of 14C-WL85871 after repeated oral dosing in the lactating cow; in-life phase and metabolite profiling (Experiment No. 6046). Unpublished report No. SBTR.93.063 from Sittingbourne Research Centre, Sittingbourne, England. Submitted to the WHO by Cyanamid, Wayne, NJ, USA.

Owen, D.E. & Butterworth, S.T.G. (1977). Toxicity of pyrethroid insecticides: Investigation of the neurotoxic potential of WL 43467 to adult domestic hens. Unpublished report No. TLGR.0134.77 from Shell Toxicology Laboratory (Tunstall). Submitted to the WHO by Cyanamid, Wayne, NJ, USA.

Pickering, R.G. (1981). A 90-day feeding study of WL 43467 in rats (Experiment No. 1806). Unpublished report No. TLGR.80.143 from Shell Toxicology Laboratory (Tunstall). Submitted to the WHO by Cyanamid, Wayne, NJ, USA.

Potter, D. & McAusland, H.E. (1980). Toxicity studies on the insecticide WL43467: A study of liver microsomal enzyme activity in rats fed WL 43467 for 2 years (Experiment No. 1105). Unpublished report from Shell Toxicology Laboratory (Tunstall). Submitted to the WHO by Cyanamid, Wayne, NJ, USA.

Price, J.B. (1984). Toxicology of pyrethroids: the acute oral and percutaneous toxicity of Ripcord (WL-43467). Technical concentrate (Project No. SRCAIR84). Unpublished report No. SBGR.84.299 from Shell Research Limited, Sittingbourne Research Centre. Submitted to the WHO by Cyanamid, Wayne, NJ, USA.

Rose, G.P. (1982a). Toxicology of pyrethroids: the acute oral and percutaneous toxicity of WL 85871 (cis-2-Ripcord) in comparison with Ripcord. Unpublished report No. SBGR.82.130 from Shell Research Limited, Sittingbourne Research Centre. Submitted to the WHO by Cyanamid, Wayne, NJ, USA.

Rose, G.P. (1982b). Toxicology of pyrethroids: the acute oral and percutaneous toxicity of WL 85871 (cis-2-RIPCORD) comparison with RIPCORD. Unpublished report No. SBGR.82.130 from Shell Research Limited, Sittingbourne Research Centre. Submitted to the WHO by Cyanamid, Wayne, NJ, USA.

Rose, G.P. (1983). Neurotoxicity of WL85871. Comparison with WL43467: The effect of twenty oral doses of WL85871 or WL43467 over a period of 4 weeks on the rat sciatic/posterior tibia nerve, trigeminal nerve and trigeminal ganglion. Unpublished report No. SBGR.83.185 from Shell Research Limited, Sittingbourne Research Centre. Submitted to the WHO by Cyanamid, Wayne, NJ, USA.

Rose, G.P. & Dewar, A.J. (1983). Intoxication with four synthetic pyrethroids fails to show any correlation between neuromuscular dysfunction and neurobiochemical abnormalities in rats. Arch. Toxicol., 53, 297-316.

Tesh, J.M., Tesh, S.A., & Davies, W. (1978). WL 43467 : Effects upon the progress and outcome of pregnancy in the rat. Unpublished report of Life Science Research. Submitted to the WHO by Cyanamid, Wayne, NJ, USA.

Tesh, J.M., Ross, F.W., & Wightman, T.J. (1984a). WL 43467: Effects of oral administration upon pregnancy in the rabbit. 1. Dosage range finding study. Unpublished test report No. 84/SHL003/014 from of Life Science Research. Submitted to the WHO by Cyanamid, Wayne, NJ, USA.

Tesh, J.M., Ross, F.W., & Wightman, T.J. (1984b). WL 43467: Effects of oral administration upon pregnancy in the rabbit. 2. Main study. Unpublished test report No. 84/SHL004/043 from Life Science Research. Submitted to the WHO by Cyanamid, Wayne, NJ, USA.

Tesh, J.M., Ross, F.W., & Wightman, T.J. (1988). Supplementary report to LSR report No. 84/SHL004/043. Unpublished report No. 88/SHL004/775 from Life Science Research. Submitted to the WHO by Cyanamid, Wayne, NJ, USA.

Thorpe, E. (1982). A 5-week feeding study with WL 85871 in rats (Experiment No. 2095). Unpublished report No. SBGR.81.212 from Shell Research Limited, Sittingbourne Research Centre. Submitted to the WHO by Cyanamid, Wayne, NJ, USA.

Thorpe, E. (1985). Fourth corrigendum/addendum to TLGR.0188.78. Unpublished report from Shell Research Limited, Sittingbourne Research Centre (Four volumes). Submitted to the WHO by Cyanamid, Wayne, NJ, USA.

Van der Bercken, J. & Vijverberg, H.P.M. (1989). Neurotoxicological effects of pyrethroid insecticides. Unpublished report from the Department of Veterinary Pharmacology, Pharmacy and Toxicology. University of Utrecht, The Netherlands. Submitted to the WHO by Cyanamid, Wayne, NJ, USA.

Vanderwaart, E.J. (1994). Evaluation of the mutagenic activity of Fastac technical in a *in vitro* mammalian cell gene mutation test with L5178Y mouse lymphoma cells (with independent repeat). Unpublished report No. 087367 from Notox B.V., The Netherlands. Submitted to the WHO by Cyanamid, Wayne, NJ, USA.

Vanderwaart, E.J. (1995). Micronucleus test in bone marrow cells of the mouse with Fastac technical. Unpublished report No. 087378 from Notox B.V., The Netherlands. Submitted to the WHO by Cyanamid, Wayne, NJ, USA.

WHO (1996). The WHO recommended classification of pesticides by hazard and guidelines to classification 1996-1997 (WHO/PCS/96.3). Available from the International Programme on Chemical Safety, World Health Organization, Geneva, Switzerland.

Wooder, M.F. (1981). Studies on the effect of WL 85871 on the integrity of rat liver DNO *in vivo* (Experiment No. 2103). Unpublished report No. SBGR.81.225 from Shell Toxicology Laboratory (Tunstall). Submitted to the WHO by Cyanamid, Wayne, NJ, USA.

ANNEXES

ANNEX 1

Reports and other documents resulting from previous meetings of the Joint FAO/WHO Expert Committee on Food Additives

1. **General principles governing the use of food additives** (First report of the Joint FAO/WHO Expert Committee on Food Additives). FAO Nutrition Meetings Report Series, No. 15, 1957; WHO Technical Report Series, No. 129, 1957 (out of print).

2. **Procedures for the testing of intentional food additives to establish their safety for use** (Second report of the Joint FAO/WHO Expert Committee on Food Additives). FAO Nutrition Meetings Report Series, No. 17, 1958; WHO Technical Report Series, No. 144, 1958 (out of print).

3. **Specifications for identity and purity of food additives (antimicrobial preservatives and antioxidants)** (Third report of the Joint FAO/WHO Expert Committee on Food Additives). These specifications were subsequently revised and published as **Specifications for identity and purity of food additives, Vol. I. Antimicrobial preservatives and antioxidants,** Rome, Food and Agriculture Organization of the United Nations, 1962 (out of print).

4. **Specifications for identity and purity of food additives (food colours)** (Fourth report of the Joint FAO/WHO Expert Committee on Food Additives). These specifications were subsequently revised and published as **Specifications for identity and purity of food additives, Vol. II. Food colours,** Rome, Food and Agriculture Organization of the United Nations, 1963 (out of print).

5. **Evaluation of the carcinogenic hazards of food additives** (Fifth report of the Joint FAO/WHO Expert Committee on Food Additives). FAO Nutrition Meetings Report Series, No. 29, 1961; WHO Technical Report Series, No. 220, 1961 (out of print).

6. **Evaluation of the toxicity of a number of antimicrobials and antioxidants** (Sixth report of the Joint FAO/WHO Expert Committee on Food Additives). FAO Nutrition Meetings Report Series, No. 31, 1962; WHO Technical Report Series, No. 228, 1962 (out of print).

7. **Specifications for the identity and purity of food additives and their toxicological evaluation: emulsifiers, stabilizers, bleaching and maturing agents** (Seventh report of the Joint FAO/WHO Expert Committee on Food Additives). FAO Nutrition Meetings Series, No. 35, 1964; WHO Technical Report Series, No. 281, 1964 (out of print).

8. **Specifications for the identity and purity of food additives and their toxicological evaluation: food colours and some antimicrobials and antioxidants** (Eighth report of the Joint FAO/WHO Expert Committee on Food Additives). FAO Nutrition Meetings Series, No. 38, 1965; WHO Technical Report Series, No. 309, 1965 (out of print).

9. Specifications for identity and purity and toxicological evaluation of some antimicrobials and antioxidants. FAO Nutrition Meetings Report Series, No. 38A, 1965; WHO/Food Add/24.65 (out of print).

10. Specifications for identity and purity and toxicological evaluation of food colours. FAO Nutrition Meetings Report Series, No. 38B, 1966; WHO/Food Add/66.25.

11. Specifications for the identity and purity of food additives and their toxicological evaluation: some antimicrobials, antioxidants, emulsifiers, stabilizers, flour-treatment agents, acids, and bases (Ninth report of the Joint FAO/WHO Expert Committee on Food Additives). FAO Nutrition Meetings Series, No. 40, 1966; WHO Technical Report Series, No. 339, 1966 (out of print).

12. Toxicological evaluation of some antimicrobials, antioxidants, emulsifiers, stabilizers, flour-treatment agents, acids, and bases. FAO Nutrition Meetings Report Series, No. 40A, B, C; WHO/Food Add/67.29.

13. Specifications for the identity and purity of food additives and their toxicological evaluation: some emulsifiers and stabilizers and certain other substances (Tenth report of the Joint FAO/WHO Expert Committee on Food Additives). FAO Nutrition Meetings Series, No. 43, 1967; WHO Technical Report Series, No. 373, 1967.

14. Specifications for the identity and purity of food additives and their toxicological evaluation: some flavouring substances and non-nutritive sweetening agents (Eleventh report of the Joint FAO/WHO Expert Committee on Food Additives). FAO Nutrition Meetings Series, No. 44, 1968; WHO Technical Report Series, No. 383, 1968.

15. Toxicological evaluation of some flavouring substances and non-nutritive sweetening agents. FAO Nutrition Meetings Report Series, No. 44A, 1968; WHO/Food Add/68.33.

16. Specifications and criteria for identity and purity of some flavouring substances and non-nutritive sweetening agents. FAO Nutrition Meetings Report Series, No. 44B, 1969; WHO/Food Add/69.31.

17. Specifications for the identity and purity of food additives and their toxicological evaluation: some antibiotics (Twelfth report of the Joint FAO/WHO Expert Committee on Food Additives). FAO Nutrition Meetings Series, No. 45, 1969; WHO Technical Report Series, No. 430, 1969.

18. Specifications for the identity and purity of some antibiotics. FAO Nutrition Meetings Series, No. 45A, 1969; WHO/Food Add/69.34.

19. Specifications for the identity and purity of food additives and their toxicological evaluation: some food colours, emulsifiers, stabilizers, anticaking agents, and certain other substances (Thirteenth report of the Joint FAO/WHO Expert Committee on Food Additives). FAO Nutrition Meetings Series, No. 46, 1970; WHO Technical Report Series, No. 445, 1970.

20. Toxicological evaluation of some food colours, emulsifiers, stabilizers, anticaking agents, and certain other substances. FAO Nutrition Meetings Report Series, No. 46A, 1970; WHO/Food Add/70.36.

21. Specifications for the identity and purity of some food colours, emulsifiers, stabilizers, anticaking agents, and certain other food additives. FAO Nutrition Meetings Report Series, No. 46B, 1970; WHO/Food Add/70.37.

22. Evaluation of food additives: specifications for the identity and purity of food additives and their toxicological evaluation: some extraction solvents and certain other substances; and a review of the technological efficacy of some antimicrobial agents. (Fourteenth report of the Joint FAO/WHO Expert Committee on Food Additives). FAO Nutrition Meetings Series, No. 48, 1971; WHO Technical Report Series, No. 462, 1971.

23. Toxicological evaluation of some extraction solvents and certain other substances. FAO Nutrition Meetings Report Series, No. 48A, 1971; WHO/Food Add/70.39.

24. Specifications for the identity and purity of some extraction solvents and certain other substances. FAO Nutrition Meetings Report Series, No. 48B, 1971; WHO/Food Add/70.40.

25. A review of the technological efficacy of some antimicrobial agents. FAO Nutrition Meetings Report Series, No. 48C, 1971; WHO/Food Add/70.41

26. Evaluation of food additives: some enzymes, modified starches, and certain other substances: Toxicological evaluations and specifications and a review of the technological efficacy of some antioxidants (Fifteenth report of the Joint FAO/WHO Expert Committee on Food Additives). FAO Nutrition Meetings Series, No. 50, 1972; WHO Technical Report Series, No. 488, 1972.

27. Toxicological evaluation of some enzymes, modified starches, and certain other substances. FAO Nutrition Meetings Report Series, No. 50A, 1972; WHO Food Additives Series, No. 1, 1972.

28. Specifications for the identity and purity of some enzymes and certain other substances. FAO Nutrition Meetings Report Series, No. 50B, 1972; WHO Food Additives Series, No. 2, 1972.

29. A review of the technological efficacy of some antioxidants and synergists. FAO Nutrition Meetings Report Series, No. 50C, 1972; WHO Food Additives Series, No. 3, 1972.

30. Evaluation of certain food additives and the contaminants mercury, lead, and cadmium (Sixteenth report of the Joint FAO/WHO Expert Committee on Food Additives). FAO Nutrition Meetings Series, No. 51, 1972; WHO Technical Report Series, No. 505, 1972, and corrigendum.

31. Evaluation of mercury, lead, cadmium and the food additives amaranth, diethylpyrocarbamate, and octyl gallate. FAO Nutrition Meetings Report Series, No. 51A, 1972; WHO Food Additives Series, No. 4, 1972.

32. **Toxicological evaluation of certain food additives with a review of general principles and of specifications** (Seventeenth report of the Joint FAO/WHO Expert Committee on Food Additives). FAO Nutrition Meetings Series, No. 53, 1974; WHO Technical Report Series, No. 539, 1974, and corrigendum (out of print).

33. **Toxicological evaluation of some food additives including anticaking agents, antimicrobials, antioxidants, emulsifiers, and thickening agents.** FAO Nutrition Meetings Report Series, No. 53A, 1974; WHO Food Additives Series, No. 5, 1974.

34. **Specifications for identity and purity of thickening agents, anticaking agents, antimicrobials, antioxidants and emulsifiers.** FAO Food and Nutrition Paper, No. 4, 1978.

35. **Evaluation of certain food additives** (Eighteenth report of the Joint FAO/WHO Expert Committee on Food Additives). FAO Nutrition Meetings Series, No. 54, 1974; WHO Technical Report Series, No. 557, 1974, and corrigendum.

36. **Toxicological evaluation of some food colours, enzymes, flavour enhancers, thickening agents, and certain other food additives.** FAO Nutrition Meetings Report Series, No. 54A, 1975; WHO Food Additives Series, No. 6, 1975.

37. **Specifications for the identity and purity of some food colours, enhancers, thickening agents, and certain food additives.** FAO Nutrition Meetings Report Series, No. 54B, 1975; WHO Food Additives Series, No. 7, 1975.

38. **Evaluation of certain food additives: some food colours, thickening agents, smoke condensates, and certain other substances.** (Nineteenth report of the Joint FAO/WHO Expert Committee on Food Additives). FAO Nutrition Meetings Series, No. 55, 1975; WHO Technical Report Series, No. 576, 1975.

39. **Toxicological evaluation of some food colours, thickening agents, and certain other substances.** FAO Nutrition Meetings Report Series, No. 55A, 1975; WHO Food Additives Series, No. 8, 1975.

40. **Specifications for the identity and purity of certain food additives.** FAO Nutrition Meetings Report Series, No. 55B, 1976; WHO Food Additives Series, No. 9, 1976.

41. **Evaluation of certain food additives** (Twentieth report of the Joint FAO/WHO Expert Committee on Food Additives). FAO Food and Nutrition Meetings Series, No. 1., 1976; WHO Technical Report Series, No. 599, 1976.

42. **Toxicological evaluation of certain food additives.** WHO Food Additives Series, No. 10, 1976.

43. **Specifications for the identity and purity of some food additives.** FAO Food and Nutrition Series, No. 1B, 1977; WHO Food Additives Series, No. 11, 1977.

44. Evaluation of certain food additives (Twenty-first report of the Joint FAO/WHO Expert Committee on Food Additives). WHO Technical Report Series, No. 617, 1978.

45. Summary of toxicological data of certain food additives. WHO Food Additives Series, No. 12, 1977.

46. Specifications for identity and purity of some food additives, including antioxidant, food colours, thickeners, and others. FAO Nutrition Meetings Report Series, No. 57, 1977.

47. Evaluation of certain food additives and contaminants (Twenty-second report of the Joint FAO/WHO Expert Committee on Food Additives). WHO Technical Report Series, No. 631, 1978.

48. Summary of toxicological data of certain food additives and contaminants. WHO Food Additives Series, No. 13, 1978.

49. Specifications for the identity and purity of certain food additives. FAO Food and Nutrition Paper, No. 7, 1978.

50. Evaluation of certain food additives (Twenty-third report of the Joint FAO/WHO Expert Committee on Food Additives). WHO Technical Report Series, No. 648, 1980, and corrigenda.

51. Toxicological evaluation of certain food additives. WHO Food Additives Series, No. 14, 1980.

52. Specifications for identity and purity of food colours, flavouring agents, and other food additives. FAO Food and Nutrition Paper, No. 12, 1979.

53. Evaluation of certain food additives (Twenty-fourth report of the Joint FAO/WHO Expert Committee on Food Additives). WHO Technical Report Series, No. 653, 1980.

54. Toxicological evaluation of certain food additives. WHO Food Additives Series, No. 15, 1980.

55. Specifications for identity and purity of food additives (sweetening agents, emulsifying agents, and other food additives). FAO Food and Nutrition Paper, No. 17, 1980.

56. Evaluation of certain food additives (Twenty-fifth report of the Joint FAO/WHO Expert Committee on Food Additives). WHO Technical Report Series, No. 669, 1981.

57. Toxicological evaluation of certain food additives. WHO Food Additives Series, No. 16, 1981.

58. Specifications for identity and purity of food additives (carrier solvents, emulsifiers and stabilizers, enzyme preparations, flavouring agents, food colours, sweetening agents, and other food additives). FAO Food and Nutrition Paper, No. 19, 1981.

59. **Evaluation of certain food additives and contaminants** (Twenty-sixth report of the Joint FAO/WHO Expert Committee on Food Additives). WHO Technical Report Series, No. 683, 1982.

60. **Toxicological evaluation of certain food additives.** WHO Food Additives Series, No. 17, 1982.

61. **Specifications for the identity and purity of certain food additives.** FAO Food and Nutrition Paper, No. 25, 1982.

62. **Evaluation of certain food additives and contaminants** (Twenty-seventh report of the Joint FAO/WHO Expert Committee on Food Additives). WHO Technical Report Series, No. 696, 1983, and corrigenda.

63. **Toxicological evaluation of certain food additives and contaminants.** WHO Food Additives Series, No. 18, 1983.

64. **Specifications for the identity and purity of certain food additives.** FAO Food and Nutrition Paper, No. 28, 1983.

65. **Guide to specifications--General notices, general methods, identification tests, test solutions, and other reference materials.** FAO Food and Nutrition Paper, No. 5, Rev. 1, 1983.

66. **Evaluation of certain food additives and contaminants** (Twenty-eighth report of the Joint FAO/WHO Expert Committee on Food Additives). WHO Technical Report Series, No. 710, 1984, and corrigendum.

67. **Toxicological evaluation of certain food additives and contaminants.** WHO Food Additives Series, No. 19, 1984.

68. **Specifications for the identity and purity of food colours.** FAO Food and Nutrition Paper, No. 31/1, 1984.

69. **Specifications for the identity and purity of food additives.** FAO Food and nutrition Paper, No. 31/2, 1984.

70. **Evaluation of certain food additives and contaminants** (Twenty-ninth report of the Joint FAO/WHO Expert Committee on Food Additives). WHO Technical Report Series, No. 733, 1986, and corrigendum.

71. **Specifications for the identity and purity of certain food additives.** FAO Food and nutrition Paper, No. 34, 1986.

72. **Toxicological evaluation of certain food additives and contaminants.** WHO Food Additives Series, No. 20. Cambridge University Press, 1987.

73. **Evaluation of certain food additives and contaminants** (Thirtieth report of the Joint FAO/WHO Expert Committee on Food Additives). WHO Technical Report Series, No. 751, 1987.

74. **Toxicological evaluation of certain food additives and contaminants.** WHO Food Additives Series, No. 21. Cambridge University Press, 1987.

75. **Specifications for the identity and purity of certain food additives.** FAO Food and Nutrition Paper, No. 37, 1986.

76. **Principles for the safety assessment of food additives and contaminants in food.** WHO Environmental Health Criteria, No. 70. Geneva, World Health Organization, 1987.

77. **Evaluation of certain food additives and contaminants** (Thirty-first report of the Joint FAO/WHO Expert Committee on Food Additives). WHO Technical Report Series, No. 759, 1987 and corrigendum.

78. **Toxicological evaluation of certain food additives.** WHO Food Additives Series, No. 22. Cambridge University Press, 1988.

79. **Specifications for the identity and purity of certain food additives.** FAO Food and Nutrition Paper, No. 38, 1988.

80. **Evaluation of certain veterinary drug residues in food** (Thirty-second report of the Joint FAO/WHO Expert Committee on Food Additives). WHO Technical Report Series, No. 763, 1988.

81. **Toxicological evaluation of certain veterinary drug residues in food.** WHO Food Additives Series, No. 23. Cambridge University Press, 1988.

82. **Residues of some veterinary drugs in animals and foods.** FAO Food and Nutrition paper, No. 41, 1988.

83. **Evaluation of certain food additives and contaminants** (Thirty-third report of the Joint FAO/WHO Expert Committee on Food Additives). WHO Technical Report Series, No. 776, 1989.

84. **Toxicological evaluation of certain food additives and contaminants.** WHO Food Additives Series, No. 24. Cambridge University Press, 1989.

85. **Evaluation of certain veterinary drug residues in food** (Thirty-fourth report of the Joint FAO/WHO Expert Committee on Food Additives). WHO Technical Report Series, No. 788, 1989.

86. **Toxicological evaluation of certain veterinary drug residues in food.** WHO Food Additives Series, No. 25, 1990.

87. **Residues of some veterinary drugs in animals and foods.** FAO Food and Nutrition Paper, No. 41/2, 1990.

88. **Evaluation of certain food additives and contaminants** (Thirty-fifth report of the Joint FAO/WHO Expert Committee on Food Additives). WHO Technical Report Series, No. 789, 1990, and corrigenda.

89. **Toxicological evaluation of certain food additives and contaminants.** WHO Food Additives Series, No. 26, 1990.

90. **Specifications for identity and purity of certain food additives.** FAO Food and Nutrition Paper, No. 49, 1990.

91. **Evaluation of certain veterinary drug residues in food** (Thirty-sixth report of the Joint FAO/WHO Expert Committee on Food Additives). WHO Technical Report Series, No. 799, 1990.

92. **Toxicological evaluation of certain veterinary drug residues in food.** WHO Food Additives Series, No. 27, 1991.

93. **Residues of some veterinary drugs in animals and foods.** FAO Food and Nutrition Paper, No. 41/3, 1991.

94. **Evaluation of certain food additives and contaminants** (Thirty-seventh report of the Joint FAO/WHO Expert Committee on Food Additives). WHO Technical Report Series, No. 806, 1991, and corrigenda.

95. **Toxicological evaluation of certain food additives and contaminants.** WHO Food Additives Series, No. 28, 1991.

96. **Compendium of Food Additive Specifications.** Joint FAO/WHO Expert Committee on Food Additives (JECFA). Combined specifications from 1st through the 37th Meetings, 1956-1990. FAO, 1992 (2 volumes).

97. **Evaluation of certain veterinary drug residues in food** (Thirty-eighth report of the Joint FAO/WHO Expert Committee on Food Additives). WHO Technical Report Series, No. 815, 1991.

98. **Toxicological evaluation of certain veterinary residues in food.** WHO Food Additives Series, No. 29, 1991.

99. **Residues of some veterinary drugs in animals and foods.** FAO Food and Nutrition Paper, No. 41/4, 1991.

100. **Guide to specifications - General notices, general analytical techniques, identification tests, test solutions, and other reference materials.** FAO Food and Nutrition Paper, No. 5, Ref. 2, 1991.

101. **Evaluation of certain food additives and naturally occurring toxicants** (Thirty-ninth report of the Joint FAO/WHO Expert Committee on Food Additives). WHO Technical Report Series No. 828, 1992.

102. **Toxicological evaluation of certain food additives and naturally occurring toxicants.** WHO Food Additive Series, No. 30, 1993.

103. **Compendium of food additive specifications: Addendum 1.** FAO Food and Nutrition Paper, No. 52, 1992.

104. **Evaluation of certain veterinary drug residues in food** (Fortieth report of the Joint FAO/WHO Expert Committee on Food Additives). WHO Technical Report Series, No. 832, 1993.

105. **Toxicological evaluation of certain veterinary drug residues in food.** WHO Food Additives Series, No. 31, 1993.

106. **Residues of some veterinary drugs in animals and foods.** FAO Food and Nutrition Paper, No. 41/5, 1993.

107. **Evaluation of certain food additives and contaminants** (Forty-first report of the Joint FAO/WHO Expert Committee on Food Additives). WHO Technical Report Series, No. 837, 1993.

108. **Toxicological evaluation of certain food additives and contaminants.** WHO Food Additives Series, No. 32, 1993.

109. **Compendium of food additive specifications, addendum 2.** FAO Food and Nutrition Paper, No. 52, Add. 2, 1993.

110. **Evaluation of certain veterinary drug residues in food** (Forty-second report of the Joint FAO/WHO Expert Committee on Food Additives). WHO Technical Report Series, No. 851, 1995.

111. **Toxicological evaluation of certain veterinary drug residues in food.** WHO Food Additives Series, No. 33, 1994.

112. **Residues of some veterinary drugs in animals and foods.** FAO Food and Nutrition Paper, No. 41/6, 1994.

113. **Evaluation of certain veterinary drug residues in food** (Forty-third report of the Joint FAO/WHO Expert Committee on Food Additives). WHO Technical Report Series, No. 855, 1995, and corrigendum.

114. **Toxicological evaluation of certain veterinary drug residues in food.** WHO Food Additives Series, No. 34, 1995.

115. **Residues of some veterinary drugs in animals and foods.** FAO Food and Nutrition Paper, No. 41/7, 1995.

116. **Evaluation of certain food additives and contaminants** (Forty-fourth report of the Joint FAO/WHO Expert Committee on Food Additives). WHO Technical Report Series, No. 859, 1995.

117. **Toxicological evaluation of certain food additives and contaminants.** WHO Food Additives Series, No. 35, 1996.

118. **Compendium of food additive specifications, addendum 3.** FAO Food and Nutrition Paper, No. 52, Add. 3, 1995.

119. **Evaluation of certain veterinary drug residues in food** (Forty-fifth report of the Joint FAO/WHO Expert Committee on Food Additives). WHO Technical Report Series, 1996.

120. **Toxicological evaluation of certain veterinary drug residues in food.** WHO Food Additives Series, No. 36, 1996.

121. **Residues of some veterinary drugs in animals and foods.** FAO Food and Nutrition Paper, No. 41/8, 1996.

122. **Evaluation of certain food additives and contaminants** (Forty-sixth report of the Joint FAO/WHO Expert Committee on Food Additives). WHO Technical Report Series, in preparation.

123. **Toxicological evaluation of certain food additives and contaminants.** WHO Food Additives Series, No. 37, in press.

124. **Compendium of food additives specifications, addendum 4.** FAO Food and Nutrition Paper, No. 52, Add. 4, in preparation.

125. **Evaluation of certain veterinary drug residues in food** (Forty-seventh report of the Joint FAO/WHO Expert Committee on Food Additives). WHO Technical Report Series, in preparation.

ANNEX 2

ABBREVIATIONS USED IN THE MONOGRAPHS

ADI	acceptable daily intake
A/G ratio	albumin/globulin ratio
ALAT	alanine aminotransferase
AP	alkaline phosphatase
ASAT	aspartate aminotransferase
bw	body weight
CNS	central nervous system
DMSO	dimethyl sulfoxide
ECG	electrocardiogram
FOB	functional observational battery
Hb	haemoglobin
HPLC	high performance liquid chromatography
i.m.	intramuscular
i.v.	intravenous
JMPR	Joint FAO/WHO Meeting on Pesticide Residues
LD_{50}	median lethal dose
MCV	mean corpuscular volume
MIC	minimal inhibitory concentration
MMAD	mass median aerodynamic diameter
MRL	maximum residue limit
NOEL	no-observed-effect level
NZW	New Zealand white (rabbit)
PCV	packed cell volume
PEG	polyethylene glycol
RBC	red blood cell
SPTN	sciatic/posterior tibial nerve
WBC	white blood cell

ANNEX 3

JOINT FAO/WHO EXPERT COMMITTEE ON FOOD ADDITIVES

Rome, 4-13 June 1996

Members

Professor L.-E. Appelgren, Professor of Pharmacology, Department of Pharmacology and Toxicology, Faculty of Veterinary Medicine, The Swedish University of Agricultural Sciences, Uppsala, Sweden

Dr D. Arnold, Deputy Director, Federal Institute for Health Protection of Consumers and Veterinary Medicine, Thielallee, Germany

Dr J. Boisseau, Director, Laboratory of Veterinary Drugs, National Centre of Veterinary and Food Studies, Fougères, France *(Chairman)*

Dr R. Ellis, Director, Chemistry Division, Food Safety and Inspection Service, US Department of Agriculture, Washington, DC, USA *(Rapporteur)*

Dr P.G. Francis, Russet House, West Horsley, Surrey, England

Dr R.D. Furrow, Beltsville, MD, USA *(Rapporteur)*

Dr Z. Hailemariam, Head, Food & Beverage Quality Control, Environmental Health Department, Ministry of Health, Addis Ababa, Ethiopia

Dr R.C. Livingston, Director, Office of New Animal Drug Evaluation, Center for Veterinary Medicine, Food and Drug Administration, Rockville, MD, USA

Dr J.D. MacNeil, Head, Food Animal Chemical Residues, Health of Animals Laboratory, Saskatoon, Saskatchewan, Canada

Dr B.L. Marshall, Counsellor Veterinary Services, New Zealand Embassy, Washington, DC, USA

Professor J.G. McLean, Pro Vice-Chancellor, Division of Science, Engineering and Design, Swinburne University of Technology, Hawthorn, Victoria, Australia *(Vice-Chairman)*

Dr A. Pintér, Deputy Director General, National Institute of Hygiene, Budapest, Hungary

Professor J. Palermo-Neto, Departamento de Patologia, Faculdade de Medicina Veterinaria e Zootecnia, Universidade de São Paulo, São Paulo, SP, Brazil

Professor A. Rico, Biochemistry-Toxicology, Physiopathology and Experimental Toxicology Laboratory (INRA), Ecole Nationale Vétérinaire, Toulouse, France[1]

Dr J.L. Rojas Martinez, Chief, Toxicology Section, National Centre for Diagnosis and Research in Animal Health, Ministry of Agriculture, San José, Costa Rica

Dr P. Sinhaseni, Associate Professor, Department of Pharmacology, Chulalongkorn University, Bangkok, Thailand

Dr S. Soback, Head, National Residue Laboratory, Kimron Veterinary Institute, Ministry of Agriculture, Beit Dagan, Israel

Dr R. Wells, Director, Research and Development, Australian Government Analytical Laboratories Pymble, Australia

Dr K. Woodward, Director of Licensing, Veterinary Medicines Directorate, Ministry of Agriculture, Fisheries and Food, Addlestone, Surrey, England

Secretariat

Dr C.E. Cerniglia, Director, Division of Microbiology, National Center for Toxicological Research, Food and Drug Administration, Jefferson, AR, USA *(WHO Temporary Adviser)*

Dr P. Chamberlain, Veterinary Medical Officer, Division of Toxicology, Office of New Animal Drug Evaluation, Center for Veterinary Medicine, Food and Drug Administration, Rockville, MD, USA *(WHO Temporary Adviser)*

Dr R. Fuchs, Head, Department of Experimental Toxicology and Ecotoxicology, Institute for Medical Research and Occupational Health, Zagreb, Croatia *(WHO Temporary Adviser)*

Dr R.J. Heitzman, Science Consultant, Newbury, Berkshire, England *(FAO Consultant)*

Dr R. Herbert, National Institute of Environmental Health Sciences, Research Triangle Park, NC, USA *(WHO Temporary Adviser)*

Dr J.L. Herrman, Assessment of Risk and Methodologies, International Programme on Chemical Safety, World Health Organization, Geneva, Switzerland *(Joint Secretary)*

Dr P.G. Jenkins, International Programme on Chemical Safety, World Health Organization, Geneva, Switzerland *(Editor)*

Dr K. Mitsumori, Chief, Third Section, Division of Pathology, Biological Safety Research Centre, National Institute of Health Sciences, Tokyo, Japan *(WHO Temporary Adviser)*

---

[1] Invited but unable to attend.

Dr J. Paakkanen, Nutrition Officer, Food and Nutrition Division, Food and Agriculture Organization of the United Nations, Rome, Italy *(Joint Secretary)*

Mrs M.E.J. Pronk, Advisory Centre of Toxicology, National Institute of Public Health and Environmental Protection, Bilthoven, The Netherlands *(WHO Temporary Adviser)*

Dr L. Ritter, Executive Director, Canadian Network of Toxicology Centres, University of Guelph, Ontario, Canada *(WHO Temporary Adviser)*

Dr G. Roberts, Director, Toxicology Evaluation Section, Commonwealth Department of Human Services and Health, Canberra, ACT, Australia *(WHO Temporary Adviser)*

Dr G.J.A. Speijers, Head of the Section Public Health of the Advisory Centre of Toxicology, National Institute of Public Health and Environmental Protection, Bilthoven, The Netherlands *(WHO Temporary Adviser)*

Dr S. Sundlof, Director, Center for Veterinary Medicine, HFV-1, Food and Drug Administration, Rockville, MD, USA *(WHO Temporary Adviser)*

ANNEX 4

Acceptable Daily Intakes, other toxicological information, and information on specifications

Adrenoceptor agonists

___

Clenbuterol

Acceptable daily intake (ADI): 0-0.004 µg per kg of body weight

Recommended maximum residue limits (MRLs)[1]

	Muscle (µg/kg)	Liver (µg/kg)	Kidney (µg/kg)	Fat (µg/kg)	Eggs (µg/kg)	Milk (µg/l)
Cattle	0.2	0.6	0.6	0.2		0.05
Horses	0.2	0.6	0.6	0.2		

[1] MRLs are expressed as the parent drug.

___

Xylazine

The Committee was unable to establish an ADI for xylazine because it concluded that a metabolite, 2,6-xylidine, is genotoxic and carcinogenic.

The Committee was unable to establish MRLs for xylazine because of the lack of information on metabolism and residue depletion in edible tissues.

The following information would be required for further review:

- Data on xylazine metabolism in target species sufficient to identify a suitable marker residue and target tissues.

- Additional data on residue depletion of xylazine and its metabolites in target species. These data should include evidence to show, in particular, whether 2,6-xylidine is present at the recommended withdrawal times.

- A suitable analytical method for determining the marker residue in target tissues.

Anthelminthic agents

Abamectin

ADI: 0-1 µg per kg of body weight[1]

Recommended maximum residue limits (MRLs)[2]

	Muscle (µg/kg)	Liver (µg/kg)	Kidney (µg/kg)	Fat (µg/kg)	Eggs (µg/kg)	Milk (µg/l)
Cattle	100	50	100			

[1] This ADI, which applies to the parent drug abamectin, was established by the 1995 Joint FAO/WHO Meeting on Pesticide Residues (JMPR; FAO Plant Production and Protection Paper 133, 1996).

[2] MRLs are expressed as avermectin B_{1a}.

Moxidectin

ADI: 0-2 µg per kg of body weight[1]

Recommended maximum residue limits (MRLs)[2]

	Muscle (µg/kg)	Liver (µg/kg)	Kidney (µg/kg)	Fat (µg/kg)	Eggs (µg/kg)	Milk (µg/l)
Cattle	20	100	50	500		
Sheep	50[3]	100	50	500		
Deer[4]	20	100	50	500		

[1] This ADI was established at the forty-fifth meeting of the Committee.

[2] MRLs are expressed as the parent drug.

[3] This MRL was established at the present meeting. All other MRLs were established at the forty-fifth meeting of the Committee. At that meeting the Committee noted the high concentration of residues at the injection site over a 35-day period after subcutaneous or intramuscular administration of the drug at the recommended dose.

[4] Temporary MRLs (see the report of the forty-fifth meeting of the Committee).

Antimicrobial agents

Chlortetracycline, oxytetracycline and tetracycline

ADI: 0-3 μg per kg of body weight[1]

Recommended maximum residue limits (MRLs)[2]

	Muscle (μg/kg)	Liver (μg/kg)	Kidney (μg/kg)	Fat (μg/kg)	Eggs (μg/kg)	Milk (μg/l)
Cattle	100	300	600			100
Pigs	100	300	600			
Sheep	100	300	600			100
Poultry	100	300	600		200	
Giant prawn (*Penaeus monodon*)	100[3]					

[1]   This ADI was established at the forty-fifth meeting of the Committee.

[2]   MRLs are expressed as the parent drug.

[3]   This MRL applies only to oxytetracycline.

Neomycin

ADI: 0-60 μg per kg of body weight

Recommended maximum residue limits (MRLs)[1]

	Muscle (μg/kg)	Liver (μg/kg)	Kidney (μg/kg)	Fat (μg/kg)	Eggs (μg/kg)	Milk (μg/l)
Cattle	500	500	10 000	500		500
Pigs	500	500	10 000	500		
Sheep	500	500	10 000	500		
Goats	500	500	10 000	500		
Chickens	500	500	10 000	500	500	
Ducks	500	500	10 000	500		
Turkeys	500	500	10 000	500		

[1]   MRLs are expressed as the parent drug.

Spiramycin

ADI: 0-50 μg per kg of body weight[1]

Recommended maximum residue limits (MRLs)

	Muscle (μg/kg)	Liver (μg/kg)	Kidney (μg/kg)	Fat (μg/kg)	Eggs (μg/kg)	Milk (μg/l)
Cattle[2]	200	600	300	300		100
Pigs[3]	200	600	300	300		
Chickens[2]	200	600	800	300		

[1] The ADI was established at the forty-third meeting of the Committee.

[2] MRLs are expressed as the sum of spiramycin and neospiramycin.

[3] MRLs are expressed as spiramycin equivalents (antimicrobially active residues).

Thiamphenicol

ADI: 0-6 μg per kg of body weight[1]

Recommended maximum residue limits (MRLs)[2]

	Muscle (μg/kg)	Liver (μg/kg)	Kidney (μg/kg)	Fat (μg/kg)	Eggs (μg/kg)	Milk (μg/l)
Cattle	40	40	40	40		
Chickens	40	40	40	40		

[1] Temporary ADI.

[2] Temporary MRLs, expressed as the parent drug.

The following information is required for evaluation in 1999:

■ Detailed reports of the carcinogenicity study in rats on which the summary report was available at the present meeting and the range-finding study used to establish dose levels in that study.

■ Residue depletion studies with radiolabelled and unlabelled thiamphenicol for identification of the marker residue and target tissues in non-ruminant cattle, chickens and pigs.

Tilmicosin

ADI: 0-40 μg per kg of body weight

Recommended maximum residue limits (MRLs)[1]

	Muscle (μg/kg)	Liver (μg/kg)	Kidney (μg/kg)	Fat (μg/kg)	Eggs (μg/kg)	Milk (μg/l)
Cattle	100	1000	300	100		
Pigs	100	1500	1000	100		
Sheep	100	1000	300	100		50[2]

[1] MRLs are expressed as the parent drug.

[2] Temporary MRL. The results of a study in lactating sheep with radiolabelled drug for estimation of the relationship between total residues and parent compound in milk are required for evaluation in 1999.

Insecticides

Cypermethrin

ADI: 0-50 μg per kg of body weight

Recommended maximum residue limits (MRLs)[1]

	Muscle (μg/kg)	Liver (μg/kg)	Kidney (μg/kg)	Fat (μg/kg)	Eggs (μg/kg)	Milk (μg/l)
Cattle	200	200	200	1000		50
Sheep	200	200	200	1000		
Chickens	200	200	200	1000	100	

[1] Temporary MRLs, expressed as the parent drug.

The following information is required for evaluation in 2000:

■ The results of radiodepletion studies that extend beyond the recommended withdrawal times using the drug in its topical formulation. The study should determine the depletion of the total residues and the parent drug in target species.

■ Evidence to verify that no interconversion of isomeric forms occurs during metabolism in the target species.

■ Further information on the validation of analytical methods, particularly data on the derivation of the limits of determination and limits of quantification.

alpha-Cypermethrin

ADI: 0-20 μg per kg of body weight

Recommended maximum residue limits (MRLs)[1]

	Muscle (μg/kg)	Liver (μg/kg)	Kidney (μg/kg)	Fat (μg/kg)	Eggs (μg/kg)	Milk (μg/l)
Cattle	100	100	100	500		25
Sheep	100	100	100	500		
Chickens	100	100	100	500	50	

[1]   Temporary MRLs, expressed as the parent drug.

The following information is required for evaluation in 2000:

■ The results of radiodepletion studies in sheep and chickens that extend beyond the recommended withdrawal times using the drug in its topical formulation. The study should determine the depletion of the total residues and the parent drug.

■ The radiodepletion study submitted for cattle should be reassessed to determine the depletion of the total residues and the parent drug.

■ Evidence to verify that no interconversion of the *cis*-isomeric forms to the *trans*-isomeric forms occurs during metabolism in the target species.

■ Further information on the validation of analytical methods, particularly data on the derivation of the limits of determination and limits of quantification.